MEASUREMENT EVALUATION

FUNDAMENTAL STUDIES IN ENGINEERING

Vol. 1 Coelho/Physics of Dielectrics for the Engineer
Vol. 2 Szücs/Similitude and Modelling
Vol. 3 Scharf, Lisieski/Amplitude Distribution Spectrometers
Vol. 4 Mohilla, Ferencz/Chemical Process Dynamics
Vol. 5 Baker et al./Explosion Hazards and Evaluation
Vol. 6 Sobczyk/Stochastic Wave Propagation
Vol. 7 König/Shakedown of Elastic-Plastic Structures
Vol. 8 Čačko, Bily, Bukoveczky/Random Processes
Vol. 9 Kubáček/Foundations of Estimation Theory
Vol. 10 Catuneanu, Mihalache/Reliability Fundamentals
Vol. 11 Boros/Measurement Evaluation

FUNDAMENTAL STUDIES IN ENGINEERING 11

MEASUREMENT EVALUATION

A. BOROS

ELSEVIER

Amsterdam · Oxford · New York · Tokyo 1989

This is a revised version of the Hungarian
Mérésértékelés
published by Műszaki Könyvkiadó, Budapest 1982

Translation by
Mrs. M. GÁBOR

Joint edition published by
Elsevier Science Publishers, Amsterdam, The Netherlands and Akadémiai Kiadó, Budapest, Hungary

The distribution of this book is being handled by the following publishers

for the U.S.A. and Canada
Elsevier Science Publishing Company, Inc.
655 Avenue of the Americas
New York, N.Y. 10010, U.S.A.

for the East European countries, Democratic People's Republic of Korea, People's Republic of Mongolia, Republic of Cuba and the Socialist Republic of Vietnam
Kultura, Hungarian Foreign Trading Company
P.O. Box 149, H—1389 Budapest 62, Hungary

for all remaining areas
Elsevier Science Publishers
25 Sara Burgerhartstraat
P.O. Box 211, 1000 AE Amsterdam, The Netherlands

Library of Congress Cataloging-in-Publication Data
Boros, Andor.
[Mérésértékelés. English]
Measurement evaluation/A. Boros.
p. cm. — (Fundamental studies in engineering; 11)
Translation of: Mérésértékelés.
Includes bibliographical references.
ISBN 0-444-98849-1
1. Mensuration. I. Title. II. Series.
T50.B6513 1989 89-17022 CIP
530.8'1—dc20

ISBN 0-444-98849-1 (Vol. 11)
ISBN 0-444-41776-7 (Series)

Printed by Akadémiai Kiadó és Nyomda Vállalat, Budapest, Hungary

CONTENTS

PREFACE

This book deals with the evaluation of measurements performed in all fields of technology and science in order to gain information about physical or technical objects. The information obtained about a measured object is called "crude" measurement information and must be related to the conditions under which the measurement took place.

Using "crude" measurement information as a starting point, evaluation is to produce physically correctly interpreted data with their estimated (or corrected) error. These can be regarded as the qualitative and quantitative properties of the measured object with the inherent (but estimated) error. Measurement and measurement evaluation do not provide absolutely accurate data but they estimate the degree of the divergence of the information from the actual value, i.e. the error of the process applied.

The book consists of three parts. The first, in three chapters, deals with the basic concepts. The second, in two chapters, treats the auxiliaries used in the evaluation of measurements. It extends to the basic concepts of probability calculus and random processes, the computing technique to be applied in evaluation, and the auxiliaries (calculators, logarithmic units, etc.).

The essential content of the book is in the four chapters of the third part entitled "Measurement Evaluation". Three of these describe the numerical and graphical evaluation of deterministic measurement information by value and by frequency, while the fourth is devoted to the fundamentals of the evaluation of random measurement information.

The content of the book and the method of presentation corresponds to my earlier monograph "Electrical Measurements in Engineering" [2] co-published in 1985 by Elsevier Science Publishers, Amsterdam, and Akadémiai Kiadó, Budapest, and may be regarded as its supplement.

Although a number of works deal with the evaluation of measurements, they either appeared a long time ago [13], [18] or serve essentially different aims [1] and [19]. This then justifies the writing and publication of this book. Doubtless, it has numerous deficiencies, but I hope that readers working in different fields of technology and engaged in making measurements of one sort or another will find it useful in solving problems, because this would disseminate the methods and enhance the interdisciplinary character of measuring techniques.

A. Boros

PART I

BASIC CONCEPTS

1. PHYSICAL QUANTITIES, UNITS, THE SI SYSTEM

The measurable properties and characteristics of physical phenomena, physical processes and/or states are called physical quantities.

Physical quantities can be defined as the product of their measures and units. Before standardization, the units to be specified in the standard may be arbitrarily chosen (preferably in a way that is easiest to reproduce) while their number provides information about the magnitude of the examined (measured, read) quantity, i.e. what fraction it is of the chosen unit, or how many times greater.

In building up systems of units for the most important or the most frequently used physical quantities, the units are arbitrarily chosen. These are called basic units, from which the so-called derived units are then formed. In all this, coherence is an important criterion. The reader will meet the concept of coherence later in this chapter.

The multiples and fractions of the basic and derived units are formed by prefixes.*

Considerable efforts are being devoted worldwide to the introduction of the internationally standardised system of units into general practice. This system has become known as the SI from the French term Système International des Unités.

Table 1.1 shows the basic units of the SI system, Table 1.2 its supplementary units.

TABLE 1.1.
Basic quantities and basic units of the SI system

Basic quantity	Basic unit, symbol
Length	metre, m
Mass	kilogram, kg
Time	second, s
Electric current	ampere, A
Thermodynamic temperature	kelvin, K
Luminous intensity	candela, cd
Amount of substance	mole, mol

* The prefixes are multiplying factors, the power of 10 concerned. For the prefixes of SI see Table 1.3.

TABLE 1.2.

The supplementary quantities and their basic units in the SI system

Supplementary quantity	Basic unit, symbol
Plane angle, angle	radian, rad
Space angle	steradian, sr

Table 1.3 shows the prefixes in the SI system together with the related decimal multipliers.

TABLE 1.3.

The prefixes of the SI system with the related decimal multipliers

Multiples			Fractions		
Decimal multiplier	Prefix	Sign	Decimal multiplier	Prefix	Sign
10^{18}	exa-	E	10^{-1}	deci-	d
10^{15}	peta-	P	10^{-2}	centi-	c
10^{12}	tera-	T	10^{-3}	milli-	m
10^{9}	giga-	G	10^{-6}	micro-	μ
10^{6}	mega-	M	10^{-9}	nano-	n
10^{3}	kilo-	k	10^{-12}	pico-	p
10^{2}	hecto-	h	10^{-15}	femto-	f
10^{1}	deca	da	10^{-18}	atto-	a

The most important rules for the use of prefixes are as follows:

1. Compound prefixes must not be used.

2. The prefix and the unit shall be regarded as a single symbol.

3. In multiples of compound fractions or products, the prefix should precede the whole unit, e.g. MN/m^2.

4. To express a quantity by number and unit it is expedient to choose a prefix which gives a number between 0.1 and 1000.

Conversion between systems of units is still required but with wider acceptance of the SI system the need will occur less frequently.

Let us now briefly survey the conversion of units, considering the concepts mentioned before (e.g. coherence). The physical quantity L, as stated, can be expressed as the product of its number $\{L\}$ and its unit $[L]$:

$$L = \{L\}[L]. \tag{1.1}$$

Example

If $L = 12$ m, then $\{L\} = 12$,

and $[L] = \text{m}$.

If the physical quantities are expressed in such a form, equations in which these quantities figure – called general quantitative equations – can be divided into an equation of numbers and an equation of units.

Examples

1. Let us take the well known quantitative equation for the peripheral speed of rotation:

$$v = D\pi n. \tag{1.2}$$

With the physical quantities v, D and n and using Eq. (1.1), Eq. (1.2) assumes the following form:

$$\{v\}[v] = \pi \{D\}[D]\{n\}[n] \tag{1.3}$$

which can obviously be divided into two equations. Introducing a factor we have the equations

$$\{v\} = \xi \{D\}\{n\}\pi \tag{1.3a}$$

and

$$[v] = \frac{1}{\xi}[D][n] \tag{1.3b}$$

where (1.3a) contains numbers only and (1.3b) contains units only. With the factor ξ we took into consideration the discrepancy in the numbers, i.e. their incoherence. If $\xi=1$, the units of the quantities in the equation are said to be coherent. In creating units, it is essential that they are made coherent. As already stated, the SI system is coherent.

In order to calculate with units, equations referred to units are defined. They are obtained by dividing the quantitative equation by a unit equation, in our case by dividing (1.2) by (1.3b):

$$\frac{v}{[v]} = \xi \frac{D}{[D]} \frac{n}{[n]} \pi. \tag{1.4}$$

2. Let us illustrate the use of equations correlated with units by the conversion of the kp force unit* of the old technical (mechanical) system into the N unit (newtons) which corresponds to it in the SI. The quantitative equation will be:

$$F = ma. \tag{1.5}$$

We may write down two referring equations, one for the technical and one for the SI system:

$$\frac{F}{[F]_m} = \xi_m \frac{m}{[m]_m} \frac{a}{[a]}, \tag{1.5a}$$

$$\frac{F}{[F]_{SI}} = \xi_{SI} \frac{m}{[m]_{SI}} \frac{a}{[a]}. \tag{1.5b}$$

* The unit kp is used in German technical literature while kgf is the unit used by the English speaking world.

Since the units in both systems are coherent in themselves, and since the units of acceleration are identical in both systems, the subscripts may be deleted. On the basis of the two previous equations we have

$$[F]_{SI} = \frac{[m]_{SI}}{[m]_m}[F]_m. \tag{1.6}$$

Substituting numerical values we have 1kg f=9.81 N.

Tables 1.4 and 1.5 serve to facilitate practical work. They show the conversion of some frequently occurring physical quantities (force, moment, pressure, work, energy, etc.) deeply ingrained in the minds of technical people, side by side with the most frequent derived SI units.

TABLE 1.4.

Conversion of old units to the present official units

Quantity	Conversion to the official unit
Length	1 Å (ångström) = 0.1 nm
Mass	1 q (quintal) = 100 kg = 0.1 t
	1 metric carat = 0.2 g
Force	1 kp (≡1 kgf) = 9.806 65 N
Pressure	1 kp/cm² (at) = 98.0665 kPa
	1 atm = 101.325 kPa
	1 Torr = $\frac{101\ 325}{760}$ Pa ≈ 133.322 Pa
	1 mmHg = 133.322 Pa = 0.133322 kPa
	1 mmH_2O = 9.80665 Pa
Mechanical stresses, strength, moduli	1 kp/mm² = 9.80665 MPa ≡ 9.80665 N/mm²
	1 kp/cm² = 98.0665 kPa ≡ 98.0665 kN/m²
Work	1 kp · m = 9.80665 J
Energy	1 erg = 0.1 μJ
Quantity of heat	1 cal (international) = 4.1868 J
Power	1 kp · m/s = 9.80665 W
	1 HP = 735.49875 W ≈ 0.735 kW
Dynamic viscosity	1 P (poise) = 0.1 Pa · s
Kinematic viscosity	1 St (stokes) = 1 cm²/s
Cross section	1 b (barn) = 100 fm²
Activity of radioactive radiation source	1 Ci (curie) = 37 Bq
Irradiation dose	1 R (roentgen) = 258 μC/kg
Dose equivalent	1 rem = 10 mSv
Absorbed dose	1 rd = 10 mGy

TABLE 1.5.

Derived units having special names

Quantity	SI unit		
	Name	Symbol	In basic and other units
Activity of radioactive source	becquerel	Bq	s^{-1}
Electric charge	coulomb	C	A · s
Capacitance	farad	F	C/V
Absorbed dose	gray	Gy	J/kg
Inductance	henry	H	Wb/A
Frequency	hertz	Hz	s^{-1}
Energy, work, heat	joule	J	N · m
Luminous flux	lumen	lm	cd · sr
Illuminance	lux	lx	lm/m^2
Force	newton	N	$kg \cdot m/s^2$
Electric resistance	ohm	Ω	V/A
Pressure, mechanical stress	pascal	Pa	N/m^2
Electric conductance	siemens	S	A/V
Dose equivalent	sievert	Sv	J/kg
Magnetic density	tesla	T	Wb/m^2
Electric potential	volt	V	W/A
Power	watt	W	J/s
Magnetic flux	weber	Wb	V · s

2. BASIC CONCEPTS OF MEASUREMENTS AND MEASURING SYSTEMS

2.1 MEASUREMENT AS A SOURCE OF INFORMATION

Physical phenomena or technical processes can be described by their physical characteristics, i.e. by the physical quantities which characterise them. Such quantities can be determined by the examination or control of the phenomenon or process concerned. The examination or control consists of establishing one or several compatible or specific properties of the object concerned. The examination may be subjective or objective.

In subjective examination, compatibility is established by the sensory organs of the examining person.

In objective examination, comparison takes place by measurements using objective devices. Objective examinations are therefore called also measurements.

Since the measurement of physical quantities is the only way to acquire reliable objective information about phenomena (processes), measurement in this context is a process in the course of which so-called measurement information is acquired. The term "measurement information" denotes the sum of all practicable information that can be acquired about a physical quantity by measurement. Measurement relates to the object concerned, and its aim is to acquire measurement information.

The measured object is a material object of unknown quantity and quality, i.e. an object whose state can be experimentally determined e.g. the level of liquid in a tank, the rotational speed of a spindle in a machine tool, the distribution of temperature in a room, etc. These are compound terms, one part of which refers to the object, the other to the physical quantity concerned.

To obtain information by measurement, a comparison is made: the number of the physical quantity examined is compared with its unit.

Comparison is an operation in which a relationship is determined between quantities of like character – notably whether one quantity is larger, equal to or smaller than the other. In practice, quantities are compared with known standards.

A measurement is direct if the comparison with the standard takes place directly, and indirect if an intermediate causal relationship is applied.

The means used for measurements are called measuring instruments or – in the general case – measuring systems. The latter consist of several individual instruments. Accordingly, a measuring system is a closed system consisting of interactive measuring instruments [2.2].

Section 2.2 will deal with the general structure and main properties of measuring systems.

2.2 STRUCTURE AND MAIN CHARACTERISTICS OF MEASURING SYSTEMS

Figure 2.1 illustrates the general layout of a measuring system. It shows several possible alternatives: on-line output, intermediate storage (e.g. magnetic tape), internal data processing, etc. Their common feature is that the sensor S of the measuring transducer MT is in direct contact with the measured object MO, and acquires information about it. This information is then converted by the transducer T into practically applicable measuring information (cf. Section 3.1). The simplest case is when the information may consist of a single datum – called a measured datum or measured value. In other cases it contains several (a set of) measured data which may be mutually dependent, or in a functional relationship to one another. Finally, in some important cases, the measurement information is carried by a continuously or discretely varying signal, called a measuring signal.

The measurement information appearing at the output of the transducer is passed to the comparing and connecting unit CC, where comparison takes place. Subsequently the signal forming unit SF converts the information obtained by comparison into a form more suitable for further processing. Following this, as the figure shows, there are alternatives for the flow of information. We do not deal with these, partly assuming that the reader is familiar with them and partly because they can be followed in the figure. A common feature of the alternatives is that the "crude" measurement information yielded by the measuring system must be evaluated in the functional unit FM. Evaluation consists of the generation of data correctly interpreted physically on the basis of the crude measurement information, with estimated or corrected errors, which can be regarded as the objective qualitative and quantitative properties of the object examined. The task of measurement evaluation is, in fact, the generation

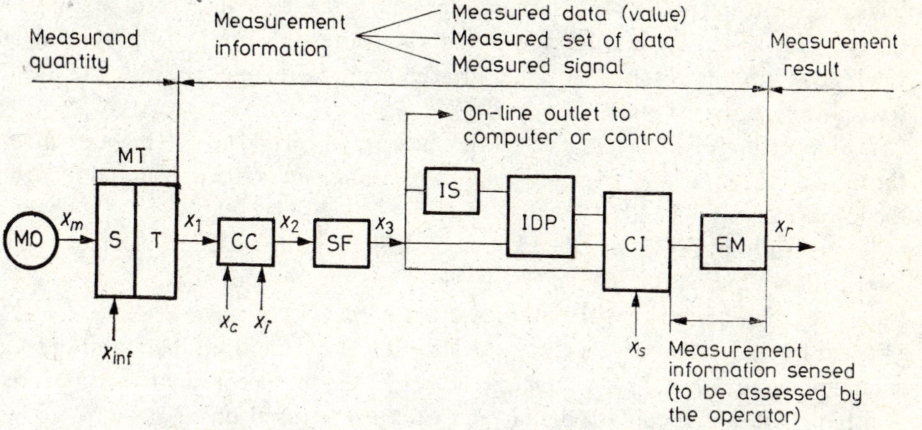

Fig. 2.1. *General layout of a measuring system*

MO—measured object, S—sensor, T—transducer, MT—measuring transducer, CC—comparing and connecting unit, SF—signal former, IS—intermediate storage (e,g. magnetic tape), IDP—internal data processing, CI—communication of information, EM—evaluation of measurement, x_m*—measured quantity,* x_{inf}*—influencing quantity,* x_1, x_2, x_3*—intermediate quantities,* x_c*—comparative quantity,* x_i*—interfering quantity,* x_s*—supplementary quantity,* x_r*—measurement result.*

of such data and the main goal of this book is to survey the means and methods available for the purpose.

Measuring systems can be classified either according to the properties of the measurement information or else according to the structural particulars of the system and its elements. In the former case the classification is information-oriented (signal-oriented), in the latter it is structure-oriented.

The structure of the measuring system is related to the path covered by the measurand quantity (for measurand see p. 14), or the measurement information, within the measuring system. Information within the system may propagate along an open chain or in a closed circle. Accordingly, open (or serial) and closed loop structures are known.

Serial (open) measuring systems generally correspond to the so-called deflection-type methods, in which the physical effect created by the measurand generates an opposite force in the measuring system. The two effects are balanced through deflection.

Closed loop measuring systems correspond to the so-called compensating methods. The essential property of the compensating processes, also known as null methods, is that they always give readings, i.e. equilibrium, in the same position, without deflection. They ensure a high degree of accuracy.

The properties described by the Standard Process Instrumentation Terminology ISA—551. 1/1976 (corresponding to MSZ 18 200 in Hungary) [2.1] relate to the functions of the measuring system. Their structural and metrological groups characterise the functions of the measuring system.

The structural characteristics include the range of signalization, the measuring range, its upper and lower limits, etc.

The signalization range is the range between the first and last graduations of the indicating instrument. The measuring range is that range of the measurand in which the error of the indicated value is less than the permissible error of the system. This value holds for the result of a single measurement carried out according to the prescribed (reference) criteria.

The upper limit of measurement is the measurand related to the highest value of the measuring range, and the lower limit of measurement is the measurand related to the lowest.

The sensor (S in Fig. 2.1) is the element of the measuring system which picks up the information required for the determination of the measurand, and passes it on. It is usually in direct contact with the measured object.

The transducer (T in Fig. 2.1) is a device suitable for the transduction of the measurand of an already converted quantity into another one, according to a known relationship. It is a self-contained unit and can also be used on its own.

The measuring transducer MT in Fig. 2.1 is the complex of sensor plus transducer.

The metrological characteristics of the measuring system are the comparative, auxiliary, intermediate, influencing, interfering parts of the system.

The comparative quantity (x_c in Fig. 2.1) is the quantity which gives the multiple or fraction of the unit with a known error, realizing the measurement directly (e.g. the length and mass of a measuring cylinder, the voltage of a normal battery, etc.)

Auxiliary quantities are the ones that are required for the measurement but cannot be obtained from the measured object.

Intermediate quantities are the ones into which the measurand or another (e.g. auxiliary) quantity is converted (e.g. force measurement by means of a strain gauge, where the intermediate quantities are the mechanical stress-strain state and the variations in electric resistance).

The influencing quantity is not the direct objective of measurement but one that exerts an effect on the measured value. Its influence can more or less be compensated or controlled.

The interfering quantity is one that exerts an unfavourable effect on the output of the measured value. Its effect cannot be controlled at all, or only with considerable difficulty. When planning measurements or a measuring system, it should be kept to the lowest possible value.

Measurements and measuring systems are also characterised by further important criteria:

— the measuring principle,
— the method of measuring and
— the measuring process.

The measuring principle is the physical phenomenon which forms the basis of the measurement (e.g. piezoelectric, capacitative, isotopic, strain-gauge, etc.).

The measuring method consists of the set of rules on the basis of which the measurement is to be carried out (direct or indirect, contact or non-contact, deflection, compensatory or differential, analogue or digital, etc.)

The measuring process is the practical application of that measuring principle and that measuring method which yield the desired information. In the course of the process the exact moment in time when a measurement was carried out is very important. Accordingly, the measure is a suitable means of indicating – once or repeatedly – one or several values of a physical quantity (e.g. monovalent measure, a measuring cylinder, weights, etc. or multivalent measures, measuring rod (scale), normal resistance decade, a set of measuring cylinders, etc.).

3. BASIC INFORMATIVE CONCEPTS OF MEASUREMENT AND MEASUREMENT EVALUATION

The author in this chapter wishes to clarify some of the concepts earlier outlined and to introduce some definitions of importance for the subsequent chapters.

3.1 MEASURAND, MEASUREMENT INFORMATION

3.1.1 BASIC DEFINITIONS

Measurand

The measurand is the physical quantity that the measurement is intended to determine. It may be represented by a single value, by several values or by a function. The real value defines the measurand under the conditions which prevail at the instant of determination, while the correct value is so close to the real value that the difference between them is negligible. The definitions of the real and the correct values can easily be generalised for cases in which the measurand is not a single value but e.g. a set of values or a function.

The measurand (physical quantity) may be constant or varying in space and/or time. (Quantities changing in space, i.e. depending on spatial coordinates, are also known as space quantities). In the general case the measurand x_m depends on place and time, and assumes the following form:

$$f(x, y, z, t) = 0, \tag{3.1}$$

$$x = x_m.$$

Since measurements are often carried out at a predetermined point in space (e.g. a point in a machine or piece of apparatus, or a point in an electrical connection), in many cases the measurand may not depend on time and the number of variables in Eq. (3.1) – which involves a function of four variables – may be reduced.

Measurement information

Figure 2.1 introduced the concept of measurement information but without its precise definition. The function (3.1) and the pertinent complementary condition $x = x_m$ contain or determine every piece of information about the measurand x_m

through its variables (dimensions, changes in time and space, relations to other quantities, etc.). Let us look at Fig. 2.1 once more. As stated before, the measuring transducer MT converts the measurand x_m into practicable measurement information, making the intermediate quantity appear at output. Thus x_{o1} has to meet only one criterion: to contain the same information as x_m. Since, as shown by (3.1), x_m contains all the information through its variables, these must agree. Accordingly, the measurement information must satisfy the following function:

$$\varphi(x, y, z, t) = 0, \qquad (3.2)$$

$$x = x_{o1}.$$

From the implicit function (3.2) the measurement information can be defined as the explicit expression for x_{o1} from (3.2), i.e. in the general case a function with three independent variables.

The function of the measurand; the measured signal

In the following, the definitions of the measurand and of the measured quantities should be treated separately because carrying the same information, they may cause a confusion.

Re-arranging Eqs (3.1) and (3.2) the explicit terms of the measurand and the measurement information are as follows:

$$x_m = \Psi(y, z, t), \qquad (3.1a)$$

$$x_{o1} = \Phi(y, z, t). \qquad (3.1b)$$

where Ψ and Φ denote the explicit functional relations corresponding to f and φ.

It has been stated that the measurand and the measured quantities carry principally the same information. How do they differ?

About the function Ψ of the measurand, we may safely say that while it describes a physical quantity in its natural – converted – state, the measured quantity which has so far been called measurement information (i.e. the quantity described by the function Φ) is its opposite and is artificially produced. To distinguish between them, the natural physical quantities will be called "functions", while the artificial physical quantities derived from the natural quantities will be called "signals" — more precisely, "measured signals". The signal, and thus also the measured signal, is the physical realisation of a physical quantity and its most important property is that it contains every piece of information relating to the changes which occur during the examination.

A signal may be every value, or change in value, of every physical state parameter (physical quantity) which is suitable for the acquisition, transfer or storage of unambiguously related information. For practical purposes, a characteristic of the selected signal carrier whose changes contain complete information on the quantity concerned is used.

Summing up: to avoid misunderstandings, x_m as expressed in (3.1a) – usually a function of y, z and t – will be called the function which expresses the measurand (in brief: "measurand function"), while x_{o1} as expressed in (3.2a) will be called "measurement information", or "measured signal".

"Measuring information" and "measured signal" are general terms.

It has been shown that the number of variables in (3.1) and naturally also in (3.2) can be reduced. By a reduction in the number of variables (as seen in 2.2) we may reduce the measurement information or the measured signal to "measured data" (measured value) or to a "measured set of data". These however are special cases and under normal conditions we shall use the terms "measurement information" and "measured signal", and make special reference to the cases where we depart from this.

3.1.2 PARAMETER OF INFORMATION. THE DIMENSIONS OF MEASUREMENT

Parameter of information

Functions (3.1) and (3.2) carry the information in their variables y, z and t. These variables are called "parameters of information", [14], [16], [21], [3.2].

The dimension of information; signal dimension

The number of informative parameters in a function which represents a physical (measurand) quantity (cf. f in (3.1)) or a measured signal (cf. 3.2) is known as the "dimension of information" or simply "signal dimension".

Since the phenomena of the material world which can be described by physical quantities take place in space and time, the information they carry can be at most $3+1=4$-dimensional (1 time and 3 space co-ordinates). In the functions (3.1) and (3.2), expressing the most general case, the information is four-dimensional. As stated, as regards the information carried the measurand and the measured signal are equivalent, so an n-dimensional signal will fit an n-dimensional measurand ($n=0$, 1, 2, 3, 4).

Now we shall examine the information carried by the measurand or the measured signal, for different values of n. For the sake of simplicity we shall speak of signals but, since the information is equivalent, the statements also hold good for the measurand.

A signal of zero dimension carries no quantitative data, the information it carries relates merely to the presence or absence of the measurand. For instance, if the state of a signal light (switched on or off) is not brought into relationship with time, it will carry qualitative information only. The examination of the state of a lamp cannot be regarded as a measurement. It is just an inspection.

The signal and the information it carries are uni-dimensional if they contain a single datum as the informative parameter — for instance when length is measured. This type of information storage is often called "linear" because it has one dimension only. Further uni-dimensional measurements are related to time, weight, current, etc. Uni-dimensional measured signals – as seen – are measured data or measured values.

On completion of a measurement, the measured value is equal to the measurand with its inherent error. In the simplest case the measurement result may be equal to the measured value. In such cases it is the value (with error) obtained on the basis of an unambiguous relationship (or identity) from one or several measured values. However, the measurement result is generally obtained by evaluation of the partial results of several measurements.

The measurement error is the discrepancy between the real or correct measured value and the measurand. There are several reasons for this.

A signal is bi-dimensional when it contains two informative parameters, i.e. when the measurand is the function of another physical quantity. For example, the measurement of the current flowing through an optional resistance as a function of voltage, or the measurement of the volume of a medium flowing across a throttle as a function of the pressure difference, etc. Bi-dimensional information is represented in the plane where the independent variable may be a spatial co-ordinate corresponding to an optional physical quantity, or time.

The cases dealt with so far, except the last one, refer to static measurements. If a quantity is a function of time, we have to deal with a dynamically varying quantity and its measurement is called "dynamic".* [23].

The signal or the information it carries is three-dimensional if it is a function of two other physical quantities (e.g. a function of two spatial or one spatial and one time quantity). The concept of the spatial quantity was explained earlier in this chapter.

Finally, the signal is four-dimensional if it is a function of three other physical quantities.

The dimensional properties of signals and of the information they carry have been compiled in Table 3.1. Using the table, every type of problem that may arise in measuring technique may be systematised.

* The term "dynamic" refers to dependence on time, but not very consistently. The only exception from this is the case of very slowly varying quantities which are called quasi-static. Stationary and non-stationarily varying quantities also exist. These terms refer to the mode of the dependence on time, and not to the dependence.

3.2 TYPES AND TASKS OF MEASURING TECHNIQUES

Here is a brief survey of the more important types of problem encountered in practice. The numbers used here are the serial numbers used in Table 3.1.

When the measured signal is zero-dimensional (1 in Table 3.1), we speak of supervision (cf. subjective examination). Static measurements are characterised by the important fact that the measurands and the measured signals corresponding to them are not time dependent. They often show no functional relations at all. Two cases may occur here: simple static measurements and time measurements.

Simple static measurement (2 in Table 3.1) takes place when the dimensions of a workpiece or the mass of an object is measured (weighed, etc.).

Time measurements (5 in Table 3.1) also fall into the category of static measurements, provided that time is in no functional relation with other physical quantities. Here it is usually the elapsed time between two events or the time period of periodic phenomena which are measured.

The next task is to plot on a graph the so-called static characteristics. Static characteristics established by measurements having one independent variable (3 in Table 3.1) — often called single-variable static characteristics — may be the measurement and plotting of e.g. current and voltage, flow and pressure difference, amplitude frequency, phase frequency, etc.

Static characteristics with two independent variables (4 in Table 3.1) are characterised by some second variables. Their plot in the plane is either parametral or nomographic (see Section 7.2.4) e.g. the characteristics of an electron tube, a transistor, etc.

Dynamic measurements are shown in the last three rows of Table 3.1.

The measurement of a simple time function (6 in Table 3.1) is among the most frequent processes in measuring. As in case 3, a plot in the plane is required, but the role of the independent variable is played here by time.

Time functions with one or two parameters (7 and 8 in Table 3.1) are less frequent. In time functions, the measurand and the measured signal depend not only on time but also on one or two spatial co-ordinates.

In practice, the types outlined above do not occur with equal frequency. Cases 1, 2 and 5 are met mainly in the field of production, in quality control or – less frequently – in the measurement of processing times. Cases 3, 4 and particularly 6, on the other hand, are met in scientific and industrial research and in development. Cases 7 and 8 are few and far-between in measuring practice.

Mention should also be made of other processes which help acquire information on different measurements. In principle there are three:

— visual observation,
— readout,
— recording.

TABLE 3.1.

The dimensional properties of measurement information (measured signals)

Number	Signal dimension	Spatial co-ordinate x, y, z	Time co-ordinate t	Mathematical form of the measured signal	Comment—example	Character of the measurement
1	0	— — —	—	—	Supervision: signal lamp on – off, gauge working – at rest	Static
2	1	* — —	—	x	Simple static measurement (of length, mass, etc.)	
3	2	** —	—	$x=f(y)$	Plotting the static characteristics with one independent variable	
4	3	***	—	$x=f(y, z)$	Plotting the static characteristics with two independent variables	
5	1	— — —	*	t	Time measurement	
6	2	* — —	*	$x=f(t)$	Measuring the time function	Dynamic
7	3	** —	*	$x=f(y, t)$	Measuring uni-parametric time function	
8	4	***	*	$z=f(y, z, t)$	Measuring bi-parametric time function	

Note: x, y, z are mostly, but not necessarily, co-ordinates of place and may also stand for physical quantities other than time (e.g. pressure, temperature, current, etc.).
Symbols: * present, — absent

Visual observation is applicable in case 1 only. It gains elementary information (e.g. that a signal lamp is on or off, a gauge is in operation or not, etc.).

Readout provides a single value of the measured signal, the one which was valid at the instant of the readout. The means may be a graduated instrument (indicator) or some other apparatus (e.g. precision-type helical potentiometer, a vernier scale, a set of gauge blocks, a digital display, etc.).

During recording the measured signal is continuously or intermittently recorded by ink or pen, using an electrical or photographic process on magnetic tape, disc or perforated type, etc. This process offers the advantage of reproducibility of the measured results and its availability at any time for documentation, etc.

When measuring functional relations with one or several variables, adjustments may become necessary with both the readout and the recording devices — adjustment meaning the "adjustment" of the independent variable(s) to the specified value(s). When the mass or the volume of a medium passing through a throttle in unit time is to be measured, as a function of the pressure difference (the independent variable), the value of the latter is generally adjusted by setting the throttle to the specified value by another throttle. The adjustment of a variable (used as parameter) maintained at a constant value during a measurement series on multivariable functions to a predetermined value (as a parameter) is also called adjustment.

In addition to measuring uni-dimensional signals, the readout method can also be used to measure signals of optional dimensions (function-like relations), but in this case several parameters must be adjusted and the functional relation can be established from point to point by independent readouts after every adjustment.

Recording can be regarded as the automated version of the readout method, in which the functional relations in a given interval are continuously recorded. If applied to bi-dimensional signals the process is illustrative (3 and 5 in Table 3.1), but when multi-dimensional signals are measured it must be remembered that the recorder will plot a continuous curve for firmly adjusted parameters only. After several adjustments the end result will be a set of parametric curves.

With regard to plots in the plane the reader's attention is drawn to Section 3.3.

3.3 TIME FUNCTIONS AND SIGNALS CORRESPONDING TO THE MEASURAND AND THE MEASUREMENT INFORMATION

We have seen that during measurement it is the measured signal, obtained from the time or other function of the measurand, reproduced by the measuring transducer and constituting the physical realisation of the quantity to be measured which carries the measurement information. It will become obvious that the task of measurement evaluation (in the wider sense of the term) is a reconstruction of the original information carried by the measurand, in a form suitable for solving practical problems, from the measured signal.

In the next Section, therefore, first the more frequently used and important time functions of the physical quantities will be described, and then the artificial signals derived from them.

3.3.1 TIME FUNCTIONS AND THEIR DESCRIPTION

While some physical quantities are independent of time i.e. constant, and are known as static quantities, some others — the majority — are time-dependent and are known as dynamic quantities.

For the exact description of dynamic quantities their time functions* are used.

1. Time functions, according to their value, may be
   — deterministic or
   — stochastic.

* The $x(t)$ time function of the physical quantity x is the analogue mathematical representation of its course in time, and carries both quantitative and qualitative information.

2. Depending on their type and their changes in time, they may be
   — stationary or
   — non-stationary.

If the change in the stationary time functions is deterministic the functions may be
— harmonic or
— periodic

while if the change is stochastic, we have
— stationary stochastic time functions.

With non-stationary time functions we also distinguish between deterministic and stochastic changes.

In the deterministic case we speak about
— transient time functions and

in the stochastic case about
— non-stationary stochastic time functions.

It should be noted that the latter are time functions of a special type, "pseudo-periodic functions" which "resemble" the periodic time functions but are non-stationary and stochastic.

A time function is deterministic if its values yield the information it carries unambiguously. It is stochastic if its course is random and the relationship between its values and the information it carries can be established by probability calculation methods only.

The change in the time function is stationary if, while it is changing, its characteristics remain constant (eg. the mean values). In non-stationary changes the characteristics of the time function also change, and we may say that the change has no "recurrent" properties.

Let us briefly survey the more important properties of time functions. Let the physical quantity be x, and its time function $x(t)$.

The instantaneous value $x(t_0)$ is the value of the time function measured at the instant t_0.

Simple (linear) mean value

The simple (linear) mean value $\bar{x}$ is the average of the time function for a test period of duration T*

$$\bar{x} = \frac{1}{T} \int_0^T x(t)\,\mathrm{d}t. \tag{3.3}$$

* In the case of periodic $x(t)$, T may be equal to the time period. For periodic time functions the integral (3.3) is interpreted as a $T \to \infty$ boundary, while in the case of non-periodic time-functions the mean obtained is time-dependent.

Effective value (root mean square)

x or $x_{\text{r.m.s.}}$, used with or without subscript. The subscript means the English abbreviation of the effective value r.m.s.=root mean square, i.e. the root mean square of the time function for the duration of the examination:

$$x = \sqrt{\overline{x^2}} = \sqrt{\frac{1}{T}\int_0^T x^2(t)\,dt}. \tag{3.4}$$

The remark made about T in (3.3) naturally also holds here. The effective value is always positive and differs from the linear mean value.

The square of the r.m.s. $\overline{x^2}=x^2$ may also be used, especially for stochastic processes, to replace the root mean square:

$$\overline{x^2} = x^2 = \lim_{T\to\infty}\frac{1}{2T}\int_{-T}^{T} x^2(t)\,dt. \tag{3.5}$$

This definition will appear later in greater detail. For the time being, we should accept that in quantities which are not constant over time $\overline{x^2}\neq\bar{x}^2$.

Absolute mean value

x_a is the mean determined from the absolute value of the time function:

$$x_a = \frac{1}{T}\int_0^T |x(t)|\,dt \tag{3.6}$$

(the remark regarding T also holds here).

Summing up: the different mean values represent different properties of the time function (e.g. the r.m.s. relates to the power represented by the time function, the absolute mean value yields the mean value of the "rectified" time function, etc.).

Now let us examine the different types of time functions and their description.

Harmonic time function

A signal consisting of a single sinusoidal or cosinusoidal oscillation is called a harmonic time function. Its general form is:

$$x(t) = \hat{x}\cos(\omega t+\varphi), \tag{3.7}$$

where $\hat{x}$ denotes the amplitude of the cosinusoidal oscillation with a phase shift of φ at an angular frequency of ω.

To describe a harmonic oscillation (3.7), the complex form

$$x(t) = \hat{x}e^{j(\omega t+\varphi)} \tag{3.8}$$

is also used, because it simplifies computations. It should be borne in mind, however, that only the real part has any physical significance. This is in agreement with (3.7) because $\mathrm{Re}\,\{x(t)\}=\hat{x}\cos(\omega t+\varphi)$. The functions (3.7) and (3.8) are periodic according to $T=1/f=2\pi/\omega$ i.e. $x(t)=x(t+nT)$, $n=0, 1, 2, \ldots$.

The characteristic values are as follows:

$$\text{amplitude } x_{\max} = \hat{x},$$

$$\text{linear mean } \bar{x} = 0,$$

$$\text{r.m.s. } x = \hat{x}/\sqrt{2}.$$

Figure 3.1 illustrates the time function (3.7) and Fig. 3.2 its "frequency spectrum" or simply "spectrum". If there is a single harmonic component, the spectrum will consist of a single line of length $\hat{x}$ and located at $\omega=2\pi/T$ (Fig. 3.2).

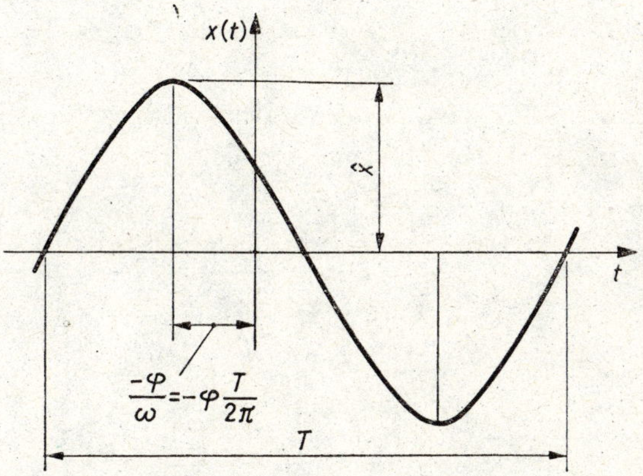

Fig. 3.1. *Harmonic time function*

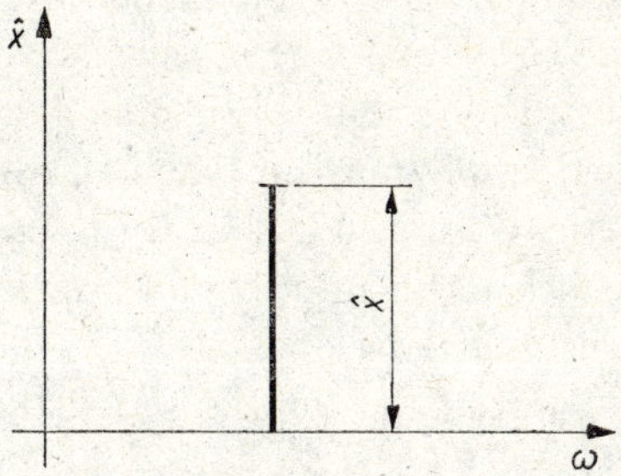

Fig. 3.2. *The frequency spectrum of a harmonic time function*

Periodic time function

The main characteristic of periodic time functions is $x(t+nT)=x(t)$ $(n=0, 1, 2, ...)$ where T stands for the period. According to the Fourier theorem, such functions can be written as the sum of harmonics whose frequencies are integral multiples of the frequency of a so-called basic harmonic. The function proper is

$$x(t) = \frac{A_0}{2} + \sum_{k=1}^{\infty} (A_k \sin k\omega t + B_k \cos k\omega t), \tag{3.9}$$

where $f = \frac{\omega}{2\pi} = \frac{1}{T}$ is the frequency of the basic harmonic.

The so-called basic frequency and the coefficients can be computed on the basis of the relationships

$$A_k = \frac{2}{T} \int_{-T/2}^{T/2} x(t) \sin k\omega t \, dt, \tag{3.10a}$$

$$B_k = \frac{2}{T} \int_{-T/2}^{T/2} x(t) \cos k\omega t \, dt. \tag{3.10b}$$

Equation (3.9) is the well known Fourier series, which can also be expressed as the sum of purely cosinusoidal members:

$$x(t) = \frac{A_0}{2} + \sum_{k=1}^{\infty} \hat{x}_k \cos (k\omega t + \varphi_k). \tag{3.11}$$

$$\hat{x} = \sqrt{A_k^2 + B_k^2} \tag{3.12a}$$

$$\varphi_k = \arctan \left[\frac{-A_k}{B_k} \right]. \tag{3.12b}$$

It should be noted, for the sake of completeness, that the value of A_0 can be calculated from (3.10a) and $\frac{A_0}{2}$ – as follows directly from (3.9) – is the mean value of the function approximated by the series and its time independent member.

Figures 3.3. and 3.4 illustrate a periodic time function and its spectrum, according to (3.11). For Figs 3.2 and 3.4 it should be noted that not only the amplitudes but also the phases can be illustrated — but this is done less frequently.

The aperiodic time function, or transient function, cannot be expanded into the (3.9) and (3.11) Fourier series. It is also customary to use the complex form of the Fourier series. We present it here because this form makes it possible to apply the Fourier method to aperiodic functions as well:

To obtain the complex form of the Fourier series, the following substitution is necessary in (3.9):

$$\cos k\omega t = \frac{e^{jk\omega t} + e^{-jk\omega t}}{2} \quad \text{and} \quad \sin k\omega t = \frac{e^{jk\omega t} - e^{-jk\omega t}}{2j}.$$

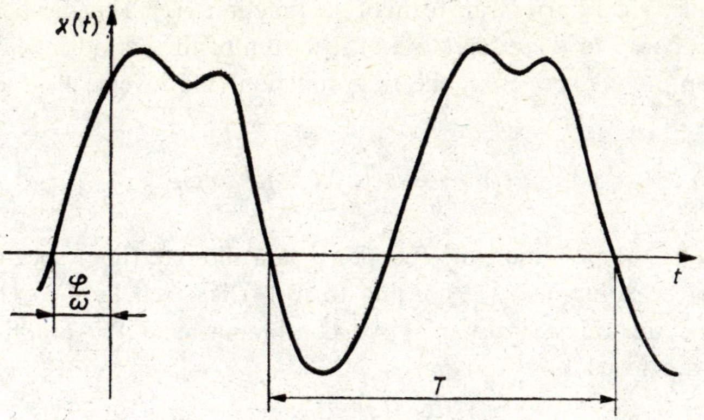

Fig. 3.3. *Periodic time function*

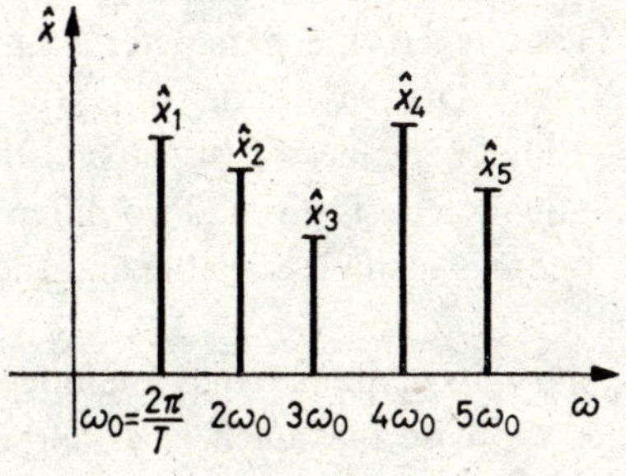

Fig. 3.4. *Line spectrum of a periodic time function*

Omitting the details of the transformations, we arrive at

$$x(t) = \sum_{k=-\infty}^{\infty} C_k e^{jk\omega t}, \tag{3.13}$$

which is in agreement with the complex form of the Fourier series. Here C_k is the complex amplitude. From the transformations (not given) we have:

$$C_k = \frac{B_k - jA_k}{2} = \frac{1}{T} \int_{-T/2}^{T/2} x(t) e^{-jk\omega t}\,dt. \tag{3.14}$$

Physical significance is assigned only to the real part of (3.13), but (3.14) can be extended to aperiodic functions without any difficulty, since they can be regarded as periodic functions with infinite time-period, i.e. $T = \infty$. Performing of (3.13) the $T \to \infty$, $\frac{1}{T} \to df$ and $k\omega \to \omega$ limes operations, again without showing the details, we obtain

$$X(j\omega) = \int_{-\infty}^{\infty} x(t) e^{-j\omega t}\,dt. \tag{3.15}$$

$X(j\omega)$ is called the Fourier transform of the function $x(t)$. This means a continuous function as opposed to the discrete (line) spectrum of the periodic case. Having obtained the continuous spectrum, the time function (the inverse Fourier transform) will be

$$x(t) = \frac{1}{2\pi} \int_{-\infty}^{\infty} X(j\omega) e^{j\omega t}\, d\omega. \tag{3.16}$$

Summing up: the periodic time function has a discrete (line) spectrum, and the amplitude of the components according to (3.14) is given by C_k. The spectra of aperiodic time functions (transients) are continuous, and the amplitude of their components are given by $X(\omega)\, d\omega$.

*Stochastic stationary time function**

This describes a stationary stochastic process, consisting of a multitude of single-parameter probability variables where the parameter t stands for time and x is a stochastically-varying quantity. The stochastic time function $x(t)$ and $x^{(1)}(t)$, $x^{(2)}(t)$ ... $x^{(i)}(t)$ denote the different values of the stochastic function $x(t)$. If all its distribution functions are known, $x(t)$ is given. Two different distribution functions of $x(t)$ are generally used, one has a single variable and the other has two variables (see Chapter 4).

Since the process is stationary, its characteristics are independent of the time of the examination. This simplifies the case inasmuch as it allows the application of the ergodic (random) hypothesis which states that instead of the stochastic functions which can never be exactly determined, its values (measured or recorded) may be used.

These values, however, would only unambiguously determine the stochastic function $x(t)$ if they were available in infinite numbers. Therefore in practice we must be satisfied by choosing simpler characteristics, even though they do not meet the demand exactly. Such characteristics are the different mean values and moments. They will be described in the next chapter. Here we give only some of the more important characteristics of the stationary stochastic function.

Simple mean value:

Ensemble mean value

$$E[x(t)] = m = \int_{-\infty}^{\infty} x f(x)\, dx. \tag{3.17}$$

Time-averaged value

$$M[x(t)] = \bar{x} = \lim_{T \to \infty} \frac{1}{T} \int_{-T}^{T} x(t)\, dt. \tag{3.18}$$

* Some of the basic concepts of probability calculus and stochastic processes are summed up and referred to in Chapter 4.

TABLE 3.2.

The characteristics and descriptions of physical quantities

The characteristic of the time function				The function of the physical quantity: Its expression	The function of the physical quantity: Form
Non-time dependent stationary (static)				$x = \hat{x} = \text{Constant}$	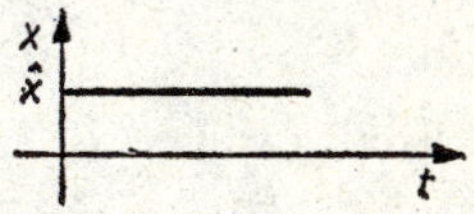
Time dependent (dynamic)	Stationary	Harmonic		$x(t) = \hat{x}\cos(\omega t + \varphi)$ Complex: $x(t) = \hat{x}e^{j(\omega t+\varphi)}$	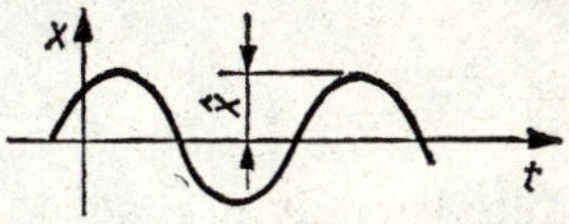
		Periodic		$x(t) = \frac{A_0}{2} + \sum_{k=1}^{\infty}(A_k \sin k\omega t + B_k \cos k\omega t)$ $x(t) = \frac{A_0}{2} + \sum_{k=1}^{\infty}\hat{x}_k \cos(k\omega t + \varphi_k)$ Complex: $x(t) = \sum_{k=-\infty}^{\infty} C_k e^{jk\omega t}$	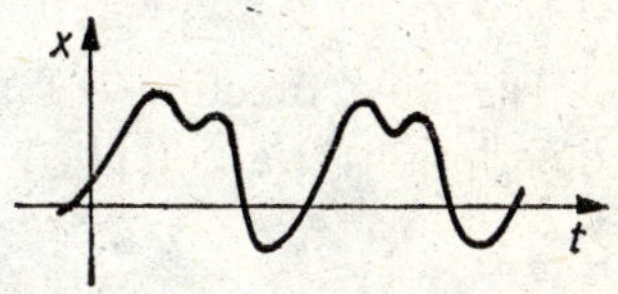
		Stochastic	Ergodic	Simple mean $\bar{x} = \lim \frac{1}{2T}\int_{-T}^{T} x(t)\,\mathrm{d}t = \int_{-\infty}^{\infty} xf(x)\,\mathrm{d}x = m$ Root mean square $x = \sqrt{\overline{x^2}}\sqrt{\int_{-\infty}^{\infty} x^2 f(x)\,\mathrm{d}x}$ Auto-correlation function $\varphi_{xx}(\tau) = E[x(t); x(t+\tau)] = M[x(t), x(t+\tau)]$	
			Non-ergodic		
	Non-stationary	Transient	Of finite energy	$x(t) = \frac{1}{2\pi}\int_{-\infty}^{\infty} X(j\omega)e^{j\omega t}\,\mathrm{d}\omega$ $X(j\omega) = \int_{-\infty}^{\infty} x(t)e^{-j\omega t}\,\mathrm{d}t$	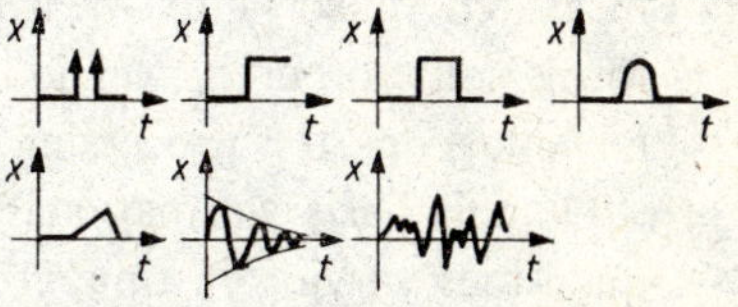
			Of non-finite energy		
		Stochastic			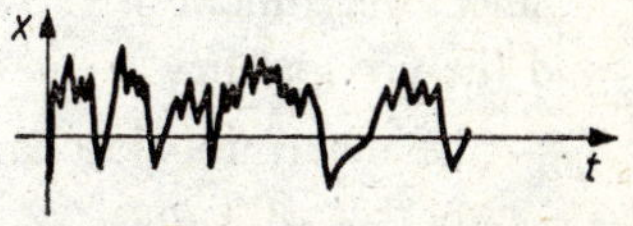
		Pseudo-periodic			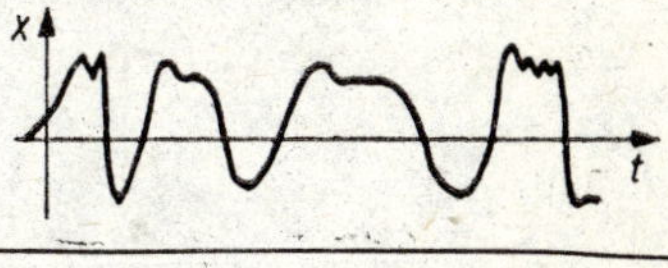

According to the ergodic hypothesis

$$E[x(t)] \equiv m = \bar{x} \equiv M[x(t)]. \tag{3.19}$$

Mean square value (r.m.s.)

Ensemble mean value

$$D[x(t)] = \sqrt{\int_{-\infty}^{\infty} x^2 f(x)\,\mathrm{d}x}. \tag{3.20}$$

Effective value

$$\sqrt{\overline{x^2}} = \sqrt{\lim_{T\to\infty} \frac{1}{T} \int_{-T/2}^{T/2} x^2 t\,\mathrm{d}t}, \tag{3.21}$$

$$D[x(t)] = \sqrt{\overline{x^2}}. \tag{3.22}$$

Auto-correlation function

$$\varphi_{xx}(\tau) = E[x(t), x(t+\tau)] = M[x(t), x(t+\tau)]. \tag{3.23}$$

The time functions of the various physical quantities shown above and their descriptions are compiled in Table 3.2.

3.3.2 SIGNALS AND THEIR DIVIDING INTO GROUPS

The signal is the physical realisation of the information carried by a physical quantity (cf. 2.2).

In Section 3.3.1 we surveyed the time functions which describe the course over time of the different measurands (physical quantities). In the simplest case, the course covered by the signal may be identical with, or similar to, the time function of the physical quantity examined, in others it is widely different from it. The only common feature of these different types of signals is the information they carry. In the latter case, i.e. while the measuring transducer transduces the quantity originating from the measured object into measurement information – measured signal – it also performs some other conversions which follow from its own working principle or are advantageous for processing the information.

On this basis we may state that while the signals which correspond to physical quantities varying in time are always time functions as well, time functions do not always appear as signals.

For more details about signals we refer the reader to the copious and easily accessible literature. However, some aspects of their classification claims attention.

Since time functions show similarity to signals, the principles of their classification and the groups they form are also similar. Signals can be divided according to

— the deterministic character of the value,
— the set of values,
— the course over time,
— the external form of the information.

According to the determination of the value the signals may be

— deterministic or
— stochastic.

The criteria are the same as those relating to time functions. According to the set of values the signals may be

— continuous (analogue) or
— intermittent (discrete).

The signal is continuous (analogue) if it can assume any value in the region examined.

The signal is intermittent (discrete) if it may assume only a finite number of defined values, and if its set of values between two adjacent discrete values is missing. According to its course over time the signal may be

— continuous (coherent) or
— interrupted (incoherent).

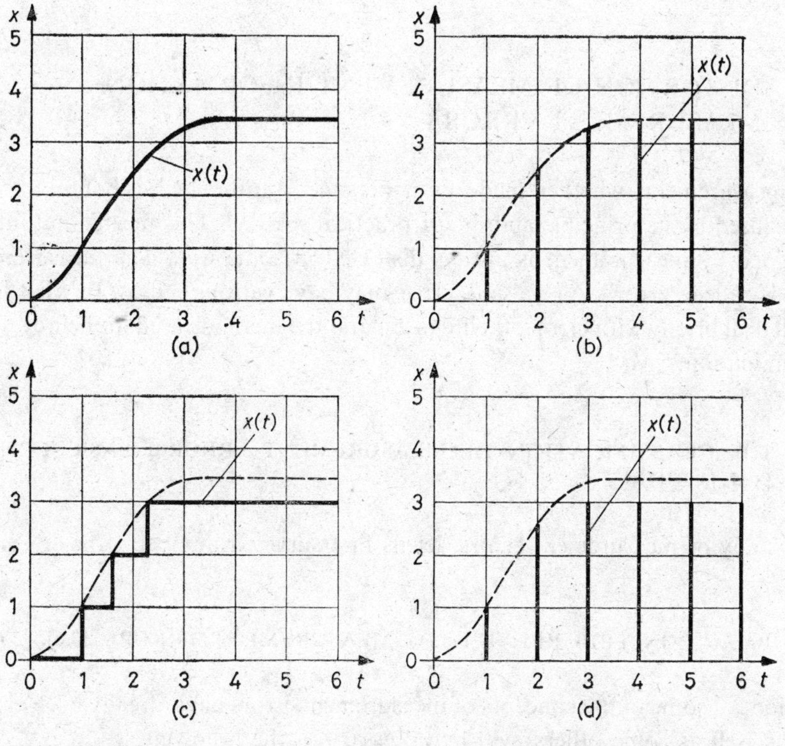

Fig. 3.5. *Different forms for the appearance of information:* (a) *x — continuous; t — flowing (analogue),* (b) *x — continuous; t — intermittent,* (c) *x — discrete; t — continuous,* (d) *x — discrete; t — intermittent (digital)*

A coherent signal remains uninterrupted for a given time period.

An interrupted signal is cut off from time to time.

Finally, according to the outward appearance of the information, similarly to the classification by the set of values, the signal may be

— analogue,
— discrete or
— digital.

An analogue signal may assume practically every value in the range examined.

A discrete signal may assume a large but finite number of values in the range examined, because its amplitude is quantized. Remember that a discrete signal may be discrete analogue or discrete digital, depending on the mode in which it carries the information.

The digital signal is a discrete signal whose values are assigned by a unified interpretation system (binary, decimal, etc.).

In Figure 3.5 (based on [15]) we show the various forms in which the information may appear, in the case of a deterministic signal.

As regards stochastic signals, we draw the reader's attention to Chapter 4.

3.4 DISTORTION OF MEASUREMENT INFORMATION, MEASUREMENT ERRORS

Every measurement, whether made by a person or automated, is bound to contain errors caused by its principle and/or for practical reasons. This means that information gained by measurement is always distorted in some way. The measure of distortion is called "error". As we said, error may have various causes. In what follows we shall deal briefly with errors, their causes, and the methods used in their evaluation and elimination.

3.4.1 THE CLASSIFICATION OF MEASUREMENT ERRORS; BASIC CONCEPTS [3.1], [3.2], [3.4]

In this study of measurement errors, let us first survey a number of basic concepts.

3.4.1.1 BASIC CONCEPTS RELATED TO MEASUREMENT ERRORS [3.5], [3.6], [3.7]

In Section 2.2 some of the concepts of measurement errors have already been clarified. They, as well as some others, will be necessary in the following.

As regards the concepts of measuring system, measurement information, signal range, range of measurement, the lower and upper measurement limits, influencing and interfering quantities, measuring principle, measuring method, measuring process etc., we draw the reader's attention to Section 2.2. The other concepts will be surveyed in the following.

The determination of the measurement range, i.e. the domain which is independent of the lower and upper limits of measurement, is of great importance. The range of measurement is that part of the scale on the instrument or system in which the measurement result is graphically displayed, the part in which measurements can be executed at a given accuracy.

The sensitivity of a measuring instrument is the differential coefficient of the angular (or linear) displacement of the pointer of the indicator instrument which corresponds to the measured quantity:

$$S = \frac{d\varphi}{dx_m}, \tag{3.24}$$

where S is the sensitivity of the measuring instrument, φ the angular displacement (shift) of the pointer of the indicator caused by a change in the value of the measured quantity and x_m is the measured quantity.

The numerical value of the resolving power of the measuring instrument is that change in the measurand which causes unit deflection. Essentially it is that minimal change in the measurand which the instrument concerned is still capable of measuring. If the function $\varphi=\varphi(x_m)$ relating the quantities in Eq. (3.24) is linear, the resolving power is equal to the sensitivity of the measuring instrument.

Accuracy is that property of the measuring instrument which establishes the difference between the measured and the real value of the measurand. Accordingly, the accuracy of a measuring instrument is characterised by its error, i.e. an opposite or complementary quantity.

After this relatively superficial definition of error, we shall go into its more precise characterization and into the classification of its types.

3.4.1.2 TYPES AND CLASSIFICATION OF MEASUREMENT ERRORS

As we have seen, every measurement carries errors, due to its principle and/or to practical causes. According to a simple but precise definition, the measurement error is the difference between the value(s) given by a measuring system, and the real value(s).

According to the accepted terminology, the real value is called the correct value, and the sensed value is known as the measured value. Since, as stated, every measurement carries some errors, the correct value cannot be determined, it has merely theoretical significance. In practice the real value is substituted for by the so-called probable or expected value, which deviates from the theoretical correct value by a negligible measure, and can therefore be regarded as being as good for reference as the real value. The probable (expected) value and the methods of its determination will be dealt with in Sections 4.1.3 and in 4.2.2.

Measurement errors can best be classified by definition. According to the basic definition, the error is the difference between a reference value and a measured value, that is

$$\Delta x = x_i - x_r, \tag{3.25}$$

where x_i is the ith measured datum and x_r the reference value to which the measured value is related, i.e. the one in relation to which the deviation (algebraic deviation) is interpreted.

If the correct value is denoted by x_0 and the expected value by $\bar{x}$ then either of the two may be the reference value. Two cases may exist:

(a) Real absolute error

$x_r = x_0$ which means that the reference value is the correct one. Substituting this into formula (3.25), the error is

$$\Delta x_0 = x_i - x_0, \tag{3.26}$$

where Δx_0 is termed the real absolute error and the reference value x_0 the correct (real) value.

(b) Virtual absolute error

$x = \bar{x}$ i.e. the reference value is the expected value. Substituting into (3.25) we have

$$\Delta \bar{x} = x_i - \bar{x}, \tag{3.27}$$

where $\Delta \bar{x}$ is the virtual absolute error.

In addition to the absolute error, which appears in the units of the measured quantity, there is another practical concept, that of relative error. By definition the relative error is the value of the absolute error related to the measurand, specifically to its reference value. According to the general concepts the relative error, denoted by δx in relation to the measured quantity x_i, on the basis of (3.25) is by definition

$$\delta x = \frac{\Delta x}{x_i} = \frac{x_i - x_r}{x_i}. \tag{3.28}$$

The x_r reference value may naturally again be the x_0 (correct) or the $\bar{x}$ (expected) value. Accordingly, the following relative types of error exist:

(c) Real relative error

$x_r = x_0$. The reference value being again the correct value. Substituting:

$$\delta x_0 = \frac{\Delta x_0}{x_0} = \frac{x_i - x_0}{x_0}, \tag{3.29}$$

where the real relative error is δx_0.

(d) Virtual relative error.

Now, derived from Eq. (3.27), we have

$$\delta\bar{x} = \frac{\Delta\bar{x}}{\bar{x}} = \frac{x_i - \bar{x}}{\bar{x}}, \tag{3.30}$$

where the virtual relative error is $\delta\bar{x}$.

Relative errors are sometimes expressed as percentages. Since dimensionless quantities are involved, we obtain the values of the relative errors of δx_0 and $\delta\bar{x}$ as percentages by multiplying the results of equations (3.29) and (3.30) by 100.

Another way to classify measuring errors is to study their causes which may be

— systematic,
— random,
— crude.

Before we go into more detail, it should be realized that the reasons for systematic errors can be determined, those for random errors cannot be determined and those for crude measurement errors may be avoided.

Systematic measurement errors may be regarded as constant. Their magnitude and sign can be determined and they can be kept under control. Their opposite is correction. Causes which can be determined include changes in the environment (temperature, auxiliary energy etc.) and constant faults of the measuring instrument used (calibration, scale graduation etc.). The characteristics of the systematic errors will be dealt with later.

The main characteristics of random errors are irregular fluctuations in magnitude and sign, which cannot be corrected by the simple methods applied in the correction of systematic errors. The reasons for their occurrence cannot be determined. They may be due to imperfect readout, uncontrollable changes in the environmental conditions (temperature, auxiliary energy), excessive play in bearings or external interference (disturbing voltage, etc.)

The estimation and determination of random errors will be the subject of Part III of this book.

Crude errors are not considered in the evaluation of measurement results. Fortunately their unusually high value never fails to attract attention. Among their – mostly avoidable – causes we find nonobservance, mistakes, faulty readout, faulty contact, sudden changes in the power supply and damage during transport. Results which arouse a suspicion of crude errors should be checked repeatedly and the measurement findings discarded.

A survey of the classification of errors would remain incomplete without mentioning that they can also be distinguished according to the type of measurand. We distinguish between

— static and
— dynamic errors.

A static error is independent of time. It occurs in the measurement of time-independent quantities with a stabilised measuring instrument. In such cases the duration of the measurement is longer than the transient time of the instrument.

With dynamic errors, the error itself is a function of time. Such errors arise in the measurement of time-dependent quantities, and in the examination and specification of the transient behaviour of measuring instruments.

What follows is a survey of the fundamentals of the systematic errors.

3.4.2 OCCURRENCE, EVALUATION AND TREATMENT OF SYSTEMATIC MEASUREMENT ERRORS

Let us study the simplified influence diagram of a measurement instrument (system) in general use. We may start with the simplification of the measuring system shown in Fig. 2.1. An instrument simplified in this way, and its main functional relations, are illustrated in Fig. 3.6, where the input signal $x_{in}=x_m$ denotes the physical quantity to be measured,

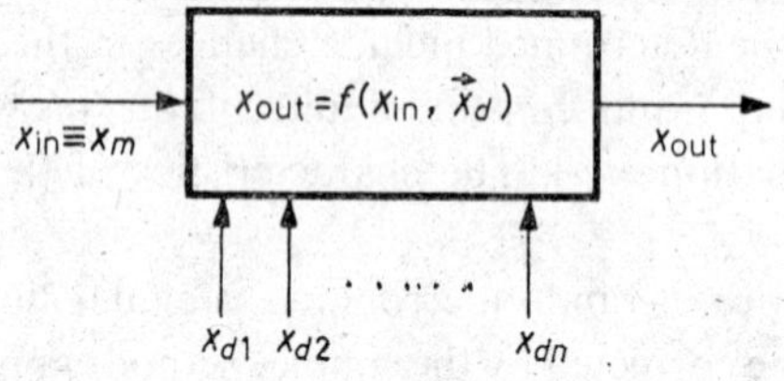

Fig. 3.6. *Simplified block diagram of a general measuring instrument (system)*

where x_{in} is the input (measurand) quantity, x_{out} the output (measured) signal and $x_{d1}...x_{dn}$ are interfering signals which may be regarded as the components of the interference vector $\vec{x}_d(x_{d1}...x_{dn})$ shown in the figure.

The $x_{in}=x_m$ measured physical quantity is distorted partly by the measuring instrument and partly by the conditions under which the measurement took place, i.e. their non-determinable effects which yield a distorted signal at its output —x_{out}.

The external conditions affect the measuring instrument by means of the interfering signals $x_{d1}...x_{dn}$ that is, by means of the interference vector $x_d(x_{d1}...x_{dn})$ they represent.

In these circumstances, the following functional relationship exists between the input and the output from the measuring instrument (see Fig. 3.6)

$$x_{out}(\equiv x_m) = f(x_{in}, \vec{x}_d). \tag{3.31}$$

It should be stated that in (3.31) x_{out} represents the measurement results containing the complete error – both the systematic and the random one (but not estimated yet), while the interfering signals $x_{d1}, x_{d2}, \ldots, x_{dn}$ and the interference vector $\vec{x}_d(x_{d1}, x_{d2}, \ldots, x_{dn})$ they stand for the interference caused by the environment. It should be remembered that the n components of the interference vector contain all interfering effects, whether determinable or non-determinable. It is expedient to define the interference vector $\vec{x}_d$ as the sum of two vector components, one of which contains the determinable and the other the non-determinable interfering effects.

If a number m of interfering effects out of n are deterministic, then in a number $(n-m)$ of cases the interfering effects cannot be traced to determinable reasons. Accordingly, the interference vector $\vec{x}_d$ can be split into two vector components, namely

$$\vec{x}_d = \vec{x}_{ds}(x_{d1}, x_{d2}, \ldots, x_{dm}) + \vec{x}_{da}(x_{dm+1}, x_{dm+2}, \ldots, x_{dn}), \tag{3.32}$$

where $\vec{x}_{ds}$ refers to determinable (systematic) reasons and $\vec{x}_{da}$ to random (non-determinable) reasons.

As said earlier we concentrate on the systematic interference vector which from (3.31) is as follows:

$$\vec{x}_{ds} = \vec{x}_{ds}(x_{d1}, x_{d2}, \ldots, x_{dm}). \tag{3.33}$$

After this introduction, let us revert to the systematic error and try to define it. Let us set out from the definition recommended by Dobrinin and Pan Čzun Čzen [3.3] which says that the absolute systematic measurement error x_{out_s} that appears at the output of the measurement instrument (system) can be regarded as the sum of the inherent systematic error $\Delta x_{out_{so}}$ and the external absolute systematic error, namely $\Delta x_{out_{ss}}$

$$\Delta x_{out_s} = \Delta x_{out_{so}} + \Delta x_{out_{ss}}. \tag{3.34}$$

We now study the typical forms of the inherent and external absolute systematic measurement errors.

3.4.2.1 THE INHERENT ERRORS OF MEASURING INSTRUMENTS

The inherent (internal) absolute systematic measuring errors $\Delta x_{out_{so}}$ (cf. (3.34)) are those caused by imperfections in the measuring instrument (system) and its deviation from the ideal. Thus they have nothing to do with the environment and the interferences. Deviation from the ideal is the deviation of the transfer characteristics to be defined at Figs 3.7 and 3.8 of the measuring instrument.

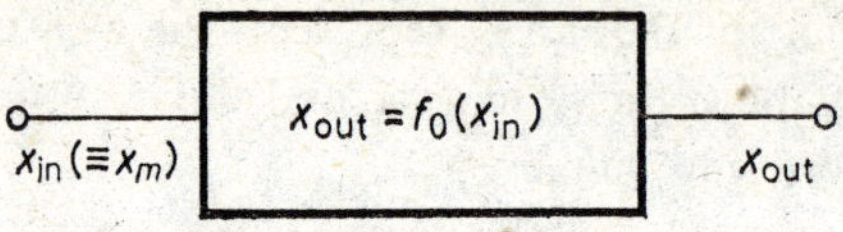

Fig. 3.7. *Idealized measuring instrument insulated from the effects of the environment*

Let us assume an ideal measuring instrument on which – for simplicity – we assume that environmental effects have been artificially eliminated. Its block diagram is shown in Fig. 3.7.

With the simplification carried out in Fig. 3.6, the functional relationship between the input and output will also become simplified:

$$x_{out} = f_0(x_{in}), \tag{3.35}$$

where f_0 refers to the inherent (internal) transfer.

Let us assume that the function f_0 between the input and the output is linear. This means that

$$x_{out} = K_0\, x_{in}, \tag{3.36}$$

where K_0 is the inherent transfer factor of the measuring instrument.

The linear transfer characteristics shown in Fig. 3.8 illustrate the linear relationship (3.36).

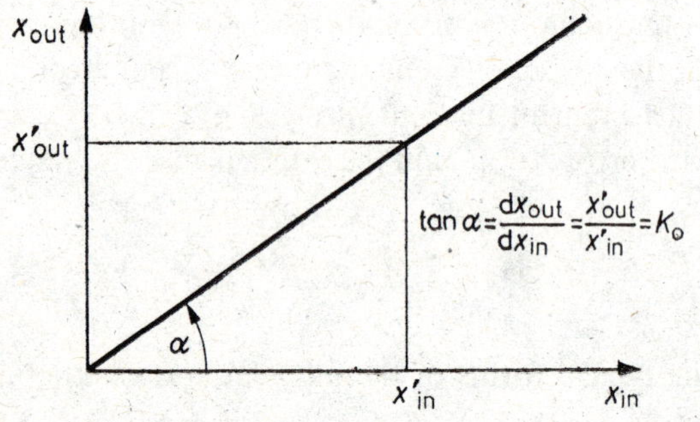

Fig. 3.8. *The linear transfer characteristic of the idealised measuring instrument illustrated in Fig. 3.7*

The error transfer of the ideal measuring instrument having the characteristics shown in Fig. 3.8 is zero. We shall revert later to the concept of error transfer.

If the characteristics shown in Fig. 3.8 – their position and/or shape – change or become distorted, the distortion will cause an internal error, caused by the measuring instrument.

The more important and typical systematic interval errors fall into three major categories:

— additive or zero errors,
— multiplicative or sensitivity errors,
— errors caused by the non-linearity of the instrument characteristics.

The additive or zero errors do not depend on x_{in}. They arise when the ideal characteristics of the measuring instrument (system) shift in parallel. Since the error caused by a parallel shift of the characteristics occurs only with linear characteristics, we shall – again for the sake of simplicity – set out once more from the ideal characteristics as shown in Fig. 3.8 (Fig. 3.9) where the ideal characteristic is shown by the dashed line. The shift may take place in both the positive and negative x_{out} senses; we shall treat the case of a positive shift.

Based on the figure, the ideal characteristics according to Eq. (3.36) can be replaced by the equation that expresses the real characteristics:

$$x_{out} = K_0 x_{in} + \Delta_a, \tag{3.37}$$

where the transfer factor K_0 is identical to the value in Eq. (3.36) and Δ_a is the additive error (zero-point error).

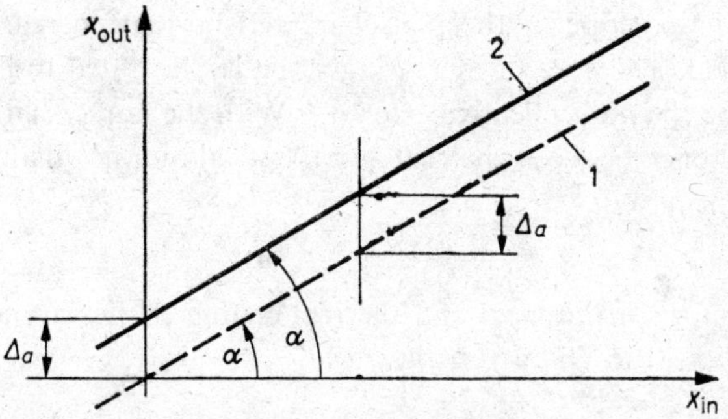

Fig. 3.9. *The occurrence of an additive (zero) error: 1 — ideal characteristics, 2 — real characteristics (containing an additive error)*
Δ_a — additive error

From Figure 3.9 and on the basis of Eq. (3.37) it will be obvious that to obtain the correct value of the measurand, the value of the additive error must be added to (or deducted from) the value of the ideal characteristic. The error to be added (or deducted) may be caused by the inherent deviation of an instrument scale, by a non-desirable permanent tare on a balance, by the thermopotential arising in d.c. measuring circuits, etc.

If ideal characteristics exist, the multiplicative or sensitivity errors will linearly depend on x_{in}. They are typically caused by a change in the directional tangent, i.e.

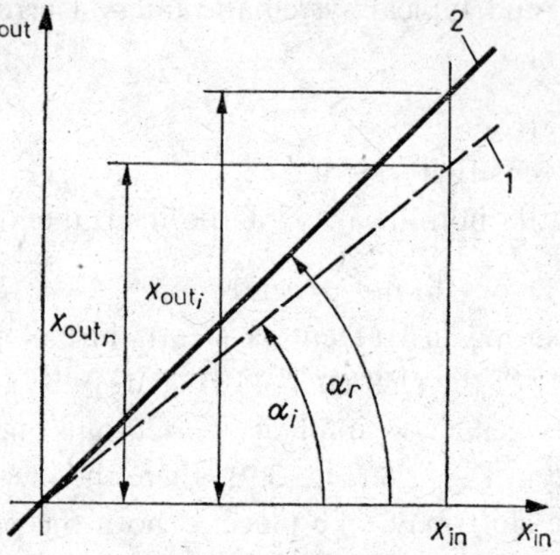

Fig. 3.10. *The occurrence of multiplicative (sensitivity) errors: 1 — ideal characteristics, 2 — real characteristics, α_i — ideal, α_r — real slope, Δ_m — multiplicative error*

the slope of the characteristic. Here, again, the change may be positive or negative which means that the slope of the real characteristic may increase or decrease in relation to the ideal. The case of increasing slope is illustrated in Fig. 3.10.

Let us determine the multiplicative error Δ_m. With the nomenclature of the figure, it will be the difference between the real and the ideal output quantities in the input (measurand) quantity x_{in}:

$$\Delta_m = x_{out_r} - x_{out_i}. \tag{3.38}$$

The transfer factors of the ideal and the real characteristics are identical to tan α of their slope. With ideal characteristics:

$$K_{0_i} = \tan \alpha_i = \frac{x_{out_i}}{x_{in}}, \tag{3.39}$$

and with real characteristics

$$K_{0_r} = \tan \alpha_r = \frac{x_{out_r}}{x_{in}}; \tag{3.40}$$

Substituting equations (3.39) and (3.40) into (3.38) we have

$$\Delta_m = (K_{0_r} - K_{0_i})\, x_{in}, \tag{3.41}$$

which shows that the multiplicative error is linearly dependent on the magnitude of the input (measurand) signal. Such an error is typically caused by the drift of measuring instruments which include electronic elements.

Additive and multiplicative errors may occur simultaneously. This case is illustrated in Fig. 3.11:

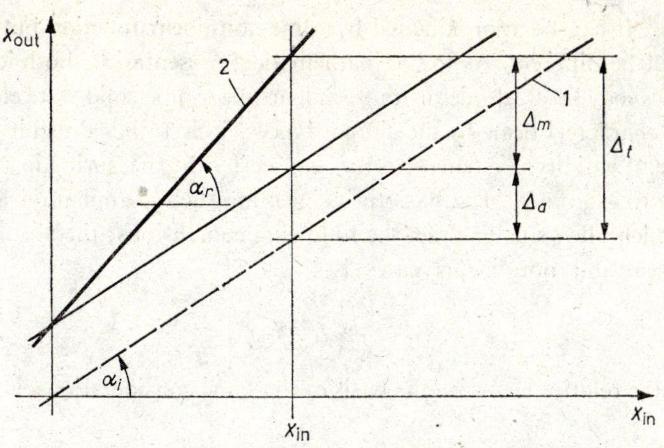

Fig. 3.11. *The combined effect of additive and multiplicative errors: 1 — ideal characteristics, — 2 real characteristics, Δ_a — additive, Δ_m — multiplicative error*

The total resultant error Δ_t is the sum of the two error components:

$$\Delta_t = \Delta_a + \Delta_m, \tag{3.42}$$

which can be determined in the manner outlined above.

The errors caused by the non-linearity of the instrument characteristics may show considerable variety, depending upon the deviation of the transfer characteristics of the instrument from the linear. In practice even an ideal characteristic may prove to

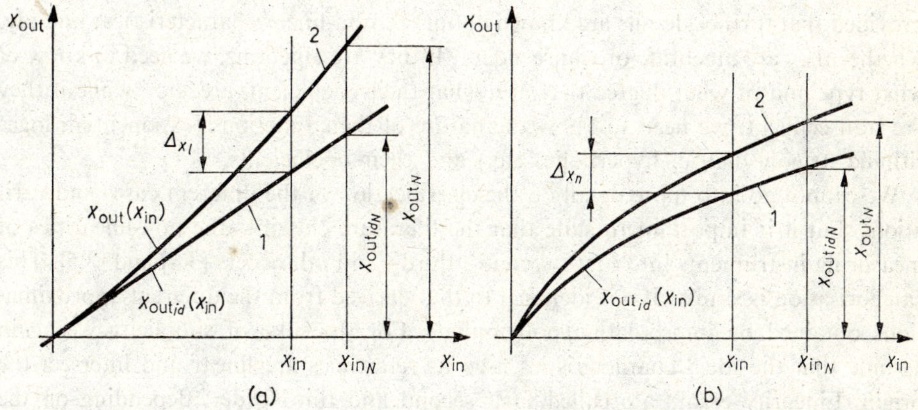

Fig. 3.12. *Errors caused by the non-linearity of the instrument in the case of* (a) *linear,* (b) *non-linear ideal characteristics: 1 — ideal, 2 — real characteristics. The subscript N refers to the nominal value, Δ_{xl} is the linear absolute error and Δ_{xn} denotes the non-linear absolute error*

be non-linear. It can be approximated by some non-linear function but its real characteristic will be different. As to the mathematical presentation, both ideal and real characteristics may be algebraic or transcendent (e.g. some solid state elements show exponential characteristics). In the circumstances, it is rather difficult to set down rules of general validity. Systematic errors caused by the measuring instrument must be individually examined. The basic principles for such examination are shown in Fig. 3.12, which allows us to draw the following conclusions: the absolute error Δx in both (linear and non-linear) cases is

$$\Delta x = x_{\text{out}} - x_{\text{out}_i}, \tag{3.43}$$

from which the relative error, on the basis of the same considerations, is by definition

$$\delta x = \frac{\Delta x}{x_{\text{out}_{id}}} = \frac{x_{\text{out}} - x_{\text{out}_{id}}}{x_{\text{out}_{id}}}. \tag{3.44}$$

Figure 3.12 shows four characteristics, three of which are non-linear. To these the inherent transfer factor defined in Eq. (3.36) applies only differentially, i.e. this factor is a function also of the value of x_{in}.

$$K_0(x_{\text{in}}) = \frac{\mathrm{d}x_{\text{out}}}{\mathrm{d}x_{\text{in}}}. \tag{3.45}$$

With the linear ideal characteristics shown in Fig. 3.12 the transfer factor is

$$K_{0_{\text{lim}}} = \frac{x_{\text{out}_{idN}}}{x_{\text{in}_{idN}}}; \tag{3.46}$$

It should be noted that from these definitions further characteristics can be derived, provided that further details are known about the non-linear characteristics: notably, whether they are algebraic or transcendent. If they are algebraic, we need to know of what type and of what degree they are, what their coefficients are, etc., while if they are transcendent, we need to know the nature of their functions (exponential, logarithmic, trigonometric, hyperbolic, etc.) and their coefficients.

We cannot go into more detail on the classification of the different cases and variations, but it is important to state that the literature classifies the non-linearities of measuring instruments into first-, second-, third-, and *n*th orders [3.4] and [3.5]. This categorisation is similar if not identical to that derived from the different approximations obtained by interpolating polynomials. For the sake of simplicity we again assume that the ideal characteristics used as references are linear and intersect the origin. Linearity errors are called first, second and third order, depending on the relation between the ideal (or presumed to be ideal) and real characteristics of the measuring instrument. Figure 3.13 illustrates in outline the typical ideal and real instrument characteristics of linearity errors of different orders.

Note: In the first-order case, the linearity error increases continuously, while in the second-order case the error first increases and then, having reached a maximum, diminishes. In the third-order case, after an initial rise the error diminishes and then changes its sign and moves towards a negative value (Fig. 3.13).

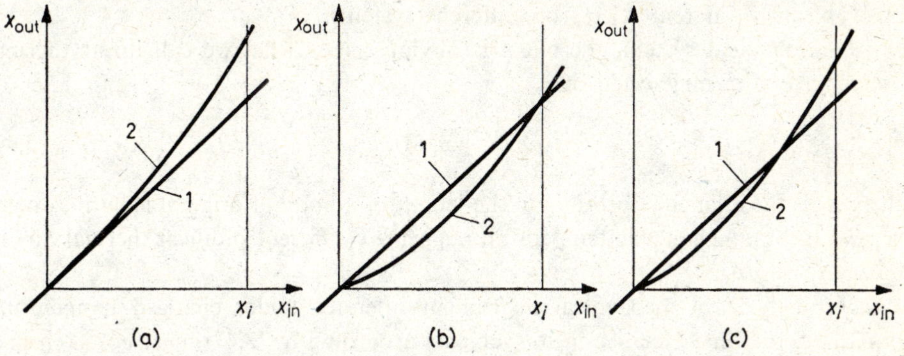

Fig. 3.13. *Linearity errors. Derivation of* (a) *first-* (b) *second- and* (c) *third-order linearity errors* *1 — ideal, linear, 2 — real measuring instrument characteristics*

3.4.2.2 INTERFERING SIGNALS CAUSED BY THE ENVIRONMENT OF THE INSTRUMENT

In Section 3.4.2 we dealt with interfering signals acting upon the system $x_{dl} \ldots x_{dn}$ and with the interference vector $\vec{x}_d(x_{d1}, x_{d2}, \ldots, x_{dn})$ derived from them, having split it into two vector components, see Eq. (3.32).

The interfering signals caused by environmental effects are due to some (deterministic or non-deterministic) change in the characteristic quantities of the environment of the measuring instrument (system). Such quantities are:

— temperature (ambient or that of the measurand),
— power supply (voltage, pressure, etc.),
— ambient humidity,
— atmospheric pressure,
— accelerations (oscillation), etc.

Depending on the type of change, these quantities cause partly deterministic variables; most of them cause random distortions in the measurement result. The case may be exemplified by two different changes in the same quantity. Let us assume that during a certain period of the day, say between 10.00 and 14.00 hours, the standard 220 V supply voltage drops regularly to 216.7 V, due to excessive industrial load. This drop and its effect can be controlled and compensated. If however the voltage drop is abrupt and stochastic, the voltage drop will get out of control and the distortion will be random. Careful deliberation is needed to determine which effect is really controllable, and its consequences must also be compensated in the measuring system. There is no general formula for such cases: the physical phenomenon (thermal, mechanical, electrical, pneumatic, etc.) and its influence must always be examined from the aspects of both cause and effect.

As a simple example, let us examine the effect of changes in the ambient temperature. This has different effects on different systems.

In a mechanical system, the effect is obvious: it is what we call linear thermal expansion of the magnitude of

$$\Delta l = \alpha_l l \Delta t, \tag{3.47}$$

where Δl is the change (algebraic) in the length of a body of original length l, due to the Δt (algebraic) change in temperature: α_l is the coefficient of linear thermal expansion.

Accordingly, if a mechanical instrument operates under changed temperature conditions, the linear change in its elements according to (3.47) must be taken into consideration. While this, in principle, determines the systematic error, the problem is sometimes rather involved.

In measuring systems containing electrical elements – resistances, semiconductors – a change in temperature will involve changes in their resistance. The same happens to metallic wires, which are not intended to function as resistors, but have low – but non-negligible – electrical resistances.

The resistance of electrical conductors and resistors is known from the well known relation:

$$R = \varrho \frac{l}{A}, \tag{3.48}$$

where R is the electrical resistance of the conductor (resistor), which depends on its geometry and the material of which it is made. The material is characterised by its specific resistance while the geometry of the conductor is characterised by its length l and "area of cross-section, A".

While thermal expansion causes only negligible changes in l and A, a change in temperature causes a considerable change in the specific resistance. Over a wide range of temperatures this change will naturally be non-linear, but in industry or in a laboratory where the range of temperatures is not very great, it is usually sufficient to rely on the first, linear member of the Taylor series which describes the temperature dependence of the change in specific resistance. In this case we obtain a relationship quite similar to (3.47):

$$\Delta \varrho = \alpha_\varrho \varrho_0 \Delta t, \tag{3.49}$$

where $\Delta \varrho$ denotes the change in specific resistance, α_ϱ its temperature factor, ϱ_0 the specific resistance in the reference state and $\Delta t = t - t_0$ the temperature change from the reference state t_0.

Such changes in resistance must always be given careful consideration in the determination of the systematic error in a measuring instrument caused by the environment. With semiconductor instruments, particularly if they are temperature-dependent like thermistors, the situation is more complicated, and examination of the properties of the materials and components is indispensable.

The other environmental effects which can be determined (supply power, humidity, atmospheric pressure, oscillations, etc.) deserve similar attention. There being no general rules to rely upon, all effects and all affected components must be considered. The simplest example is the manometer, which measures pressure, compared with atmospheric pressure as a reference value. If the atmospheric pressure changes (e.g. due to changes in elevation above sea level), this must be compensated for by an additive (algebraic) correction.

Similar considerations are required with regard to other environmental effects.

3.4.3 RANDOM MEASUREMENT ERRORS

The factors which cause random measurement errors cannot be determined, and their effects and magnitude are not controllable by simple means. For their determination or correction, the methods of probability calculus are applied. These will be the subject of the second and third parts of this book.

PART II

AIDS USED IN MEASUREMENT EVALUATION

4. PROBABILITY CALCULUS, STOCHASTIC (RANDOM) PROCESSES (INTRODUCTION)

When random errors and their consequences are examined in measurement evaluation, some important probability and allied concepts must be applied in choosing the type of approximation to be used and in measuring non-deterministically varying quantities. What follows is a summary of these concepts to the depth necessary to approach the problems dealt with in this book. We wish to call the reader's attention to the fact that our treatment does not replace a systematic study of probability theory: it is meant merely to survey the subject.

No derivations or verifications are presented, only some principles.

4.1 PROBABILITY CALCULUS

4.1.1 BASIC CONCEPTS

Random test

The problems of probability calculus relate to a so-called "random test", the results of which are not unambiguously determined by the conditions considered. (The term "test" here has its more general meaning and is not confined to the artificial production of phenomena.)

Let us consider a random test which may yield a number n of possible results, called "elementary events".

If, regarding an elementary event A, the most favourable case is denoted by k ($k \leqq n$) then the probability of the occurrence of the event A will be:

$$P(A) = \frac{k}{n} = \frac{\text{number of favourable events*}}{\text{number of possible events}} . \tag{4.1}$$

* The relationship (4.1) is the classic definition of probability. Since this is not a detailed textbook on probability theory and since the space available to us is limited, it seemed preferable to use the above relationship instead of Kolmogorov's system of axioms which requires an extensive apparatus. For Kolmogorov's postulates see References [3], [20], [4.1] and [4.2].

Accordingly, probability is nothing but a numerical value linked to the event concerned, in the way outlined above. If the event cannot possibly occur, then $k=0$ and thus $P(A)=0$. If the event A is certain to occur, then $k=n$ and $P(A)=1$. Since generally $k<n$, $0 \leqq P(A) \leqq 1$, and the probability is a value between 0 and 1.
Example: What is the probability of throwing a 6 with a die? The number of possibilities in $n=6$, the number of favourable cases

$$k = 1; \text{ thus } \quad P = \frac{k}{n} = \frac{1}{6}.$$

Example: What is the probability of throwing an even number? The number of possible cases is $n=6$, the number of favourable cases is 3 (namely, 2, 4, 6);

$$\text{thus} \quad P = \frac{k}{n} = \frac{3}{6} = \frac{1}{2}.$$

Basic relationships of probabilities

As seen before, the probability of impossible events is 0, and that of events which will certainly occur, is 1.

Events $A_1, A_2, \ldots, A_n$ will create a system of events if one and only one of them is certain to occur. In a system of events the sum of the probability of the individual elementary events yields 1:

$$P(A_1)+P(A_2)+\ldots+P(A_n) = 1. \tag{4.2}$$

For any pair of events A and B we have

$$P(A+B) = P(A)+P(B)-P(AB)^*. \tag{4.3}$$

In accordance with (4.3) for the events $A_1, A_2, \ldots, A_n$

$$P(A_1+A_2+\ldots+A_n) = P(A_1)+P(A_2)+\ldots+P(A_n). \tag{4.4}$$

Conditional probability

Let two events be A and B and let us carry out the experiment related to them. Let the number of cases in which event B occurs be k_B. This includes a number k_{AB} of events in which A, too, occurs (together with B). The quotient of the two frequencies of occurrence will be k_{AB}/k_B which seems to remain constant even if the number of experiments is increased. Since

$$\frac{k_{AB}}{n} = P(AB) \tag{4.5}$$

* $A+B$ in the algebra of events (Boolean algebra) denotes the *OR* relationship between events A and B. This means that either A or B or both will occur. AB, on the other hand, denotes the *AND* relationship, i.e. the simultaneous (coincidental) occurrence of A and B.

and

$$\frac{k_B}{n} = P(B) \tag{4.6}$$

it follows that

$$\frac{k_{AB}/n}{k_B/n} = \frac{P(AB)}{P(B)} P(A|B). \tag{4.7}$$

where $P(A|B)$ is the conditional probability of the occurrence of the event A referred to the event B.

Independence

Two events, A and B, are stochastically independent if

$$P(A|B) = P(A), \tag{4.8}$$

which, according to the definition of conditional probability in (4.7), satisfies the criterion of

$$P(AB) = P(A)P(B). \tag{4.9}$$

The concept of independence, as will be seen, is indispensable in the treatment of numerous practical problems.

4.1.2 RANDOM VARIABLES AND THEIR DISTRIBUTIONS

Random variables

A random variable is a function evaluated from a set of elementary events. The distribution function $F(x)$ of the ξ random variable establishes the probability of ξ assuming values less than x:

$$F(x) = P(\xi < x). \tag{4.10}$$

This function, the so-called (integral) distribution function increases monotonically from $F(-\infty)=0$ to $F(+\infty)=+1$ and is always continuous from the left, i.e. $\lim_{x \to x_0 - 0} F(x) = F(x_0)$.

Types of distribution

There are two basic types of random distribution:

— discrete,
— continuous.

The properties of the random variable and the distribution function described above hold for both discrete and continuous distributions.

Discrete distribution

In the case of a discrete distribution the possible values of the ξ probability (random) variable may form a finite or an infinite series. Let the probabilities of the possible values $x_1, x_2, ..., x_i$ be $p_1, p_2, ..., p_i$, then

$$p_i = P(\xi = x_i), \quad i = 1, 2, 3, ... \tag{4.11}$$

and the distribution function:

$$F(x) = P(\xi < x) = \sum_{x_i < x} p_i. \tag{4.12}$$

Continuous distribution

The distribution is continuous if there is a function $f(x) \geqq 0$ which in a definite interval (x_1, x_2) meets the criterion of

$$F(x_2) - F(x_1) = P(x_1 \leqq \xi < x_2) < \int_{x_1}^{x_2} f(x)\,dx. \tag{4.13}$$

In the case of a continuous and differentiable distribution function the so-called probability-density function is

$$f(x) = F'(x) \tag{4.14}$$

i.e. the differential coefficient of the distribution function.

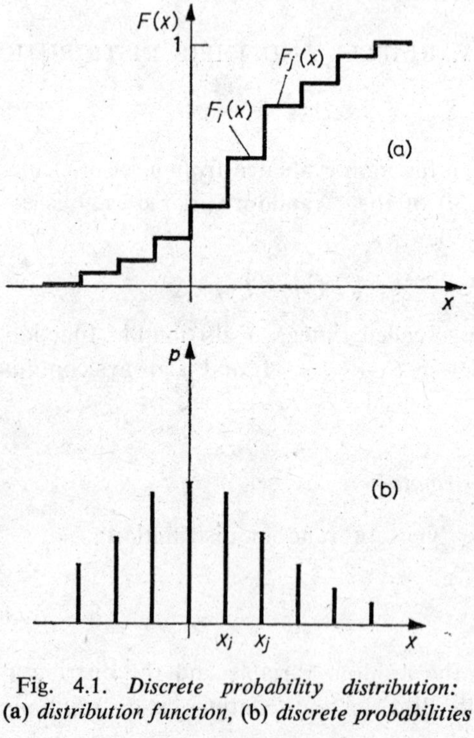

Fig. 4.1. *Discrete probability distribution:*
(a) *distribution function,* (b) *discrete probabilities*

The values of the density function are not negative and they meet the criterion which follows from the properties of $F(x)$:

$$\int_{-\infty}^{\infty} f(x)\,dx = 1. \tag{4.15}$$

With a discrete distribution, the density function cannot be interpreted because $F(x)$ cannot be continuously differentiated. Notwithstanding, if we have a sufficient number of data (e.g. measured data), the discrete probabilities can be approximated fairly well by a continuous function.

In Figs 4.1 and 4.2 discrete and continuous distribution functions with their density functions are illustrated. The figures show another important property:

$$P(x_i \leqq \xi < x_j) = F(x_i) - F(x_j), \tag{4.16a}$$

$$P(x \leqq \xi < x + dx) = \int_{x}^{x+dx} f(x)\,dx. \tag{4.16b}$$

As regards (4.16b), we present a histogram which, in the case of a continuous distribution, allows a practicable approximation to the density function. In the case of

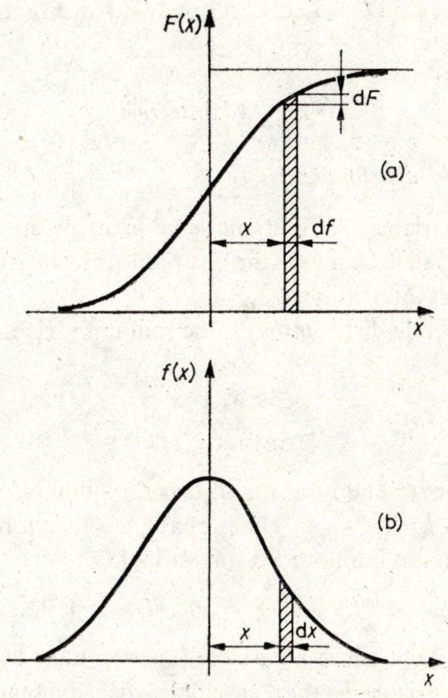

Fig. 4.2. *Continuous probability distribution:* (a) *distribution function,* (b) *density function*

a continuous distribution the interval which contains the possible values of ξ is divided into a large number of smaller intervals of length Δx and the test related to ξ is executed n times. Let k_x be the number of values below x, and draw a rectangle of height $\dfrac{k_{x+\Delta x}-k_x}{n\Delta x}$ above the $x+\Delta x$ interval.

The height is the relative density of the event $x \leqq \xi \leqq x+\Delta x$ divided by Δx. The boundary value of the sum of the areas of the rectangles is in agreement with the integral (4.16b), while its upper sides approximate to a density function. The histogram is shown in Fig. 4.3.

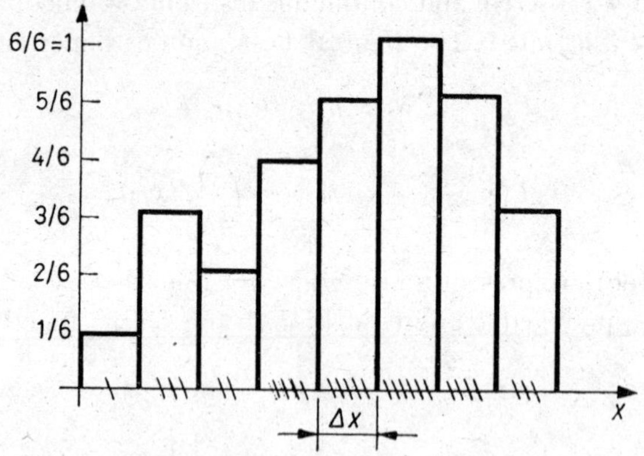

Fig. 4.3. *Histogram*

Multivariable distributions

The so-called multivariable distributions have great practical significance. For the sake of simplicity, we shall deal with the bivariable case, but the findings can be generalised to any number of variables.

In the case of a discrete distribution, if the random variables are ξ and η, their possible values will be

$$\xi: x_1, x_2, \ldots$$

$$\eta: y_1, y_2, \ldots$$

and the possible values (ξ and η) of the so-called random vector variable lie between the vector (x_i, y_k), $(i, k=1, 2, \ldots)$. The probabilities therefore determine the discrete distribution of the (ξ, η) random vector variable

$$p_{ik} = P(\xi = x, \eta = y) \quad (i, k = 1, 2, \ldots). \tag{4.17}$$

In a continuous distribution a bivariable function must exist which, $F(x, y)$ as in the case of a single variable, is the "ensemble" density function $f(x, y) \geqq 0$ of the "ensemble" distribution of the two variables. Its definition is:

$$F(x, y) = P(\xi < x, \eta < y). \tag{4.18}$$

Between the two functions the relationship

$$f(x, y) = \frac{\partial^2 F(x, y)}{\partial x \, \partial y} \tag{4.19}$$

exists.

4.1.3 CHARACTERISTICS OF RANDOM DISTRIBUTION

For the characterization of random distributions, different numerical data are used of which the expected value and the scatter are of greatest importance.

The expected value

The expected value is the centre, as it were, of the random variable distribution. Since during a test series the assumed values of the random variables fluctuate around a defined value, the arithmetic mean of the assumed values also fluctuates around it. Thus the aforementioned centre, which is a theoretical value, is called the "expected value" of the random variable.

If, in the discrete case, ξ assumes the values x_k $(k=1, 2, \ldots)$ with a probability of p_k $(k=1, 2, \ldots)$ then the expected value of ξ denoted by $E(\xi)$ will be

$$E(\xi) = \sum_{k=1}^{\infty} p_k x_k. \tag{4.20}$$

If ξ may assume an infinite number of values, then the expected value by the definition given in (4.20) is the absolute convergence of the line under the symbol Σ only, and can be evaluated as $\sum_{k=1}^{\infty} |x_k| p_k < \infty$.

In the continuous case, if ξ is a continuous random variable whose density function is $f(x)$, then the expected value of ξ will be

$$E(\xi) = \int_{-\infty}^{\infty} x f(x) \, \mathrm{d}x. \tag{4.21}$$

As in the discrete case, a secondary criterion is also necessary here. This criterion is the absolute integrability, which means that

$$\int_{-\infty}^{\infty} |x| f(x) \, \mathrm{d}x < \infty$$

because otherwise the expected value cannot be evaluated.

Since both (4.20) and (4.21) are similar to the formulae for the moments of plane figures calculated about the axis, $E(\xi)$ is known also as the first-order moment of the random distribution.

$E(\xi)$ has some important properties:

1. $E(c\xi) = cE(\xi)$ (if c = const.). (4.22a)

2. $E(\xi_1+\xi_2+\ldots+\xi_n) = E(\xi_1)+E(\xi_2)+\ldots+E(\xi_n)$, (4.22b)

3. $E(\xi+c) = E(\xi)+c$, (4.22c)

provided that $\xi_1=\xi$ and $\xi_2=c$

4. $E(\xi_1, \xi_2) = E(\xi_1)\,E(\xi_2)$, 4.22d)

provided that ξ_1 and ξ are statistically independent.

Scatter

The scatter measures the fluctuation of the random variable around the expected value. Since the momentary deviation of ξ from the expected value may be positive or negative, and since the measure of fluctuation should be independent of sign, for the scatter the positive square root of the variance is used.

The variance is defined as the expected value of the square of the difference between ξ and $E(\xi)$

$$D^2(\xi) = E\{[\xi-E(\xi)]^2\}, \tag{4.23a}$$

and the scatter is

$$D(\xi) = +\sqrt{E\{[\xi-E(\xi)]^2\}}. \tag{4.23b}$$

After squaring and using the relationship (4.22a) we arrive at

$$D^2(\xi) = E(\xi^2)-2E^2(\xi)+E^2(\xi) = E(\xi^2)-E^2(\xi) \tag{4.24}$$

which says that the variance can be calculated as a difference between the expected value of the square of the random variable and the square of the expected value.

The variance can be determined from the discrete probabilities that is, the continuous distribution function.

In the discrete case (using (4.20)):

$$D^2(\xi) = \sum_k [x_k-E(\xi)]^2 p_k, \tag{4.25}$$

in the continuous case (using (4.21))

$$D^2(\xi) = \int_{-\infty}^{\infty} [x-E(x)]^2 f(x)\,\mathrm{d}x, \tag{4.26}$$

or, transformed according to (4.24)

$$D^2(\xi) = \int_{-\infty}^{\infty} x^2 f(x)\,\mathrm{d}x - E^2(\xi) = \mu_2-\mu_1^2. \tag{4.27}$$

Here $\mu_1=E(\xi)$ is the first-order moment of the distribution while μ_2 denotes the integral in (4.27) which is the second-order moment of the density function for the case when $x=0$.

The scatter has two significant properties

1. If ξ has a scatter, a and b are two real constants and we can prove that

$$D^2(a\xi+b) = a^2D^2(\xi). \tag{4.28}$$

2. If $\xi_1, \xi_2 \ldots \xi_n$ are independent probability variables and do have a scatter then their sums will also have scatter and give:

$$D^2(\xi_1+\xi_2+\ldots+\xi_n) = D^2(\xi_1)+D^2(\xi_2)+\ldots+D^2(\xi_n). \tag{4.29}$$

In the same way as for the first- and second-order moments, higher order moments can also be defined. They yield valuable information about the nature of the density function. For instance, with a continuous distribution, in the nth order case,

$$\mu_n = \int_{-\infty}^{\infty} x^n f(x)\,\mathrm{d}x. \tag{4.30}$$

Median

Although generally different from the expected value, the median is also suitable for the estimation of the expected absolute deviation of the scatter from a given value. If b denotes this given value it is expedient to choose a numerical value for which

$$E(|\xi-b|) = \text{is minimal!} \tag{4.31}$$

This b is called the median. The practical definition of the median is possible on the basis of the distribution function. If there is a solution to the equation

$$F(x) = \frac{1}{2}, \tag{4.32}$$

where the distribution function is $F(x)$ then this solution is the median of the distribution. Note that Eq. (4.32) can be resolved in three different ways (see Fig. 4.4).

In Fig. 4.4a the equation is solved by a single value of x i.e. to the median denoted by b. It is sometimes (mainly in German literature) denoted by $\tilde{x}$.

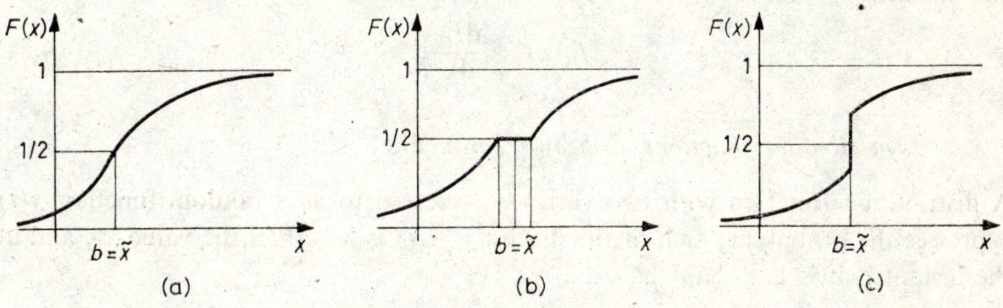

Fig. 4.4. *Interpretation of the median*

In Fig. 4.4b the solutions of Eq. (4.32) cover a whole interval, whose centre is called the median.

Figure 4.4c occurs in discrete and mixed distributions. Here the distribution function jumps at the ordinate value 1/2. In such cases the value of x, where the jump takes place, is called the median.

4.2 STOCHASTIC (RANDOM) PROCESSES

4.2.1 BASIC CONCEPTS

Stochastic processes are the single-parameter probability variables of ξ_t where the parameter t stands for time.

If the random quantity x varies as a function of time, i.e. $x = x(t)$ $(-\infty < t < \infty)$ then during each observation signal courses $x^{(1)}(t)$, $x^{(2)}(t)$, ... will take place. These are called the realisations of the random quantity $x(t)$, or the stochastic (random) signal.

The random function $x(t)$ is unambiguously defined if each of its distribution functions are known. If the distribution functions are continuous and differentiable then it is sufficient to know the density functions of $x(t)$ only.

Distribution functions

Distribution functions express the probability that at certain instant(s) the random function $x(t)$ is less than a defined value of x. Since functions with one and two variables are of greatest significance, we shall concentrate on them exclusively.

Distribution functions with one variable

A distribution function with a single variable expresses the probability that at the instant t_1 $x(t)$ is less than a defined value, i.e.

$$F_{t1}(x) = P[x(t_1) < x], \tag{4.33}$$

and its density function:

$$f_{t1}(x) = \frac{\mathrm{d}F_{t1}}{\mathrm{d}x}. \tag{4.34}$$

Distribution functions with two variables

A distribution function with two variables, referred to as a random function $x(t)$ expresses the probability that at the instant t_1 $x(t)$ is less than the value x_1, and at the instant t_2 it is less than the value x_2, i.e.

$$F_{t1,t2}(x_1, x_2) = P[x(t_1) < x_1, x(t_2) < x_2], \tag{4.35}$$

whose density function is as follows:

$$f_{t1,t2}(x_1, x_2) = \frac{\partial^2 F_{t1,t2}}{\partial x_1 \partial x_2}. \tag{4.36}$$

Similarly, a distribution function with two variables may express the distribution of two random probability variables (stochastic quantity) when common distribution is concerned. The distribution function with two variables

$$F_{t1,t2}(x, y) = P[x(t_1) < x, y(t_2) < y] \tag{4.37}$$

expresses the probability that at the instant t_1 the random function $x(t)$ is less than x and at the instant t_2 the random function $y(t)$ is less than y. The density function with two variables corresponding to (4.37) is as follows:

$$f_{t1\ t2}(x, y) = \frac{\partial^2 F_{t1,t2}}{\partial x \partial y}. \tag{4.38}$$

Stationary random processes

It is characteristic of these processes that their distribution and density functions do not depend on the instant at which the examination takes place. Accordingly, it is irrelevant whether they are determined at the instant t_k or $t_k+\tau$.

In the case with a single variable

$$F_{t1}(x) = F_{t1+\tau}(x), \tag{4.39a}$$

$$f_{t1}(x) = f_{t1+\tau}(x). \tag{4.39b}$$

With two variables, considering (4.35) and (4.36)

$$F_{t1,t2}(x_1, x_2) = F_{t1+\tau,t2+\tau}(x_1, x_2), \tag{4.40a}$$

$$f_{t1,t2}(x_1, x_2) = f_{t1+\tau,t2+\tau}(x_1, x_2), \tag{4.40b}$$

and similar relationships can be written down for (4.37) and (4.38).

Equations (4.39) and (4.40) show that for stationary random processes the distribution functions with a single variable are independent of time and that the distribution functions with two variables are independent of the instants t_1 and t_2 of examination. They depend solely on t_2-t_1 – to be denoted in the following by τ. $\tau = t_2 - t_1$.

4.2.2 MEAN VALUE AND THE CORRELATION FUNCTION

In random processes the random function $x(t)$ is usually unknown, and cannot be directly determined by measurements either. Since however the values $x^{(1)}(t), x^{(2)}(t), \ldots$ $\ldots, x^{(i)}(t)$ of the function $x(t)$ are measurable and also suitable for recording, we may assume that some of these are known to us. In the practice of measurements it is not advisable to consider too many values because they carry excessive amounts of information which are difficult to use. It is preferable to choose characteristics which, although they do not allow us to determine the random process unambiguously, yield more useful information. Such characteristics are e.g. the moments of the random function.

The crudest characteristic is the first-order moment or mean value, more precisely the expected value. (For its definition see (4.21).)

Substituting the density function from (4.34):

$$\mu_1(t_1) = E[x(t_1)] = \int_{-\infty}^{\infty} x f_{t1}(x)\,\mathrm{d}x. \tag{4.41}$$

A more precise characterization can be obtained from the second-order moment or the correlation function, which is the expected value of the product of the values of the random function measured at two different instants of time. Thus, substituting the density function (as in (4.36)) we have

$$\mu_{11}(t_1, t_2) = E[x(t_1), x(t_2)] = \int_{-\infty}^{\infty}\int_{-\infty}^{\infty} x_1 x_2 f_{t1,t2}(x_1, x_2)\,\mathrm{d}x_1\,\mathrm{d}x_2. \tag{4.42}$$

As seen, in a stationary process $f_{t1}(x)$ is not dependent on t_1 and $f_{t1,t2}(x_1, x_2)$ is not dependent on t_1 and t_2. It depends solely on the difference in time $\tau = t_1 - t_2$. Thus the expected value (mean value (4.41)) is independent of time, and the correlation function depends only on:

$$\mu = m = E[x(t)], \tag{4.43}$$

$$\mu_{11} = \varphi(\tau) = E[x(t), x(t+\tau)], \tag{4.44}$$

where m denotes the mean value, also called the mean value of the ensemble and φ denotes the correlation function.

The expected value is also called the mean value because, as we have seen, it marks the centre as it were of the random variable. The mean value can, of course, also be calculated in other ways. According to the concept it may be expected that the mean value of the stationary random function $x(t)$ will be its average over an infinitely long time. Let the mean value so obtained be denoted by M

$$M[x(t)] = \bar{x} = \lim_{T\to\infty} \frac{1}{2T} \int_{-T}^{T} x(t)\,\mathrm{d}t. \tag{4.45}$$

The mean value of the function $x(t)x(t+\tau)$ i.e. the correlation function, can be produced in a similar manner, by averaging:

$$\overline{x(t)x(t+\tau)} = M[x(t)x(t+\tau)] = \lim_{T\to\infty} \frac{1}{2T} \int_{-T}^{T} x(t)x(t+\tau)\,\mathrm{d}t. \tag{4.46}$$

The application of the concepts of the mean value and the correlation function simplifies the description of the stationary random process. But the problem still remains, that the random function $x(t)$ is unknown, although knowledge of it would be necessary for the computation of the mean values, correlation functions, etc. in our relationships. Since measurements yield only some $x^{(i)}(t)$ realisations of $x(t)$ we must find ways to arrive at the random $x(t)$ and find out how to compute the previously defined characteristics on the basis of measurement information alone.

The solution to the problem lies in the so-called ergodic theorem, or ergodic hypothesis, which states that provided that $x(t)$ is a stationary random function which meets certain requirements, the mean value computed in two different ways – as the expected value and as the average over time — is the same. The same also holds true for the correlation functions calculated in two different ways.

According to the ergodic hypothesis:

$$m \equiv E[x(t)] = M[x(t)] \equiv \bar{x}, \tag{4.47}$$

$$\varphi(\tau) \equiv E[x(t)x(t+\tau)] = M[x(t)x(t+\tau)] \equiv \overline{x(t)x(t+\tau)}. \tag{4.48}$$

It also follows from the ergodic hypothesis, that instead of the random function averaging may be aimed at a single value of $x^{(i)}-(t)$ by considering the probability meaning of the random boundary value:

$$\bar{x} = \lim_{T\to\infty} \frac{1}{2T} \int_{-T}^{T} x^{(i)}(t)\,\mathrm{d}t, \tag{4.49}$$

$$\varphi(\tau) = \lim_{T\to\infty} \frac{1}{2T} \int^{T} x^{(i)}(t)x^{(i)}(t+\tau)\,\mathrm{d}t. \tag{4.50}$$

This property is confirmed by the justified assumption that the recordings are statistically equivalent as well as by the ergodic hypothesis.

A stationary random process is expected to yield the information sought by the examination of a single signal (value) during an appropriately long period of time.

Auto- and cross-correlation functions

The functions defined by (4.42) and (4.44) have so far simply been called correlation functions, although they are frequently termed autocorrelation functions, because they are derived from the distribution function (4.35) and the density function (4.36). They should be distinguished from (4.37) and (4.38), which are derived from the distribution and density functions in a similar manner, but are known as cross-correlation functions. The following brief summary will explain the meaning of the two terms.

The auto-correlation function – as we have seen – is the mean value (moment) of some stationary random function $x(t)$ which, according to the ergodic hypothesis, can be reached in two ways. For clarity, we shall use the notation $\varphi_{xx}(\tau)$ which gives:

$$\varphi_{xx}(\tau) = E[x(t), x(t+\tau)] = M[x(t), x(t+\tau)]. \tag{4.51}$$

Also the cross-correlation function can be defined for two stationary random functions, $x(t)$ and $y(t)$ or, according to the ergodic hypothesis, for their simultaneous arbitrary recordings $x^{(i)}t$ and $y^{(i)}(t)$ in a way similar to that for the auto-correlation function in (4.42), with the only difference that instead of (4.36) the (4.38) density function is used. Without detailed calculations, the cross correlation function denoted by $\varphi_{xy}(\tau)$ is

$$\varphi_{xy}(\tau) = E[x(t), y(t+\tau)] = M[x(t), y(t+\tau)]. \tag{4.52}$$

The auto- and cross-correlation functions lend considerable help to the description of randomly varying physical quantities. However, we restrict our treatment to their most essential properties only.

The auto-correlation function is an even function, i.e.

$$\varphi_{xx}(\tau) = \varphi_{xx}(-\tau), \tag{4.53}$$

which assumes its highest value at $\tau=0$. It is important that in this case it yields the r.m.s. value, i.e. the effective value, of the stationary random function:

$$\varphi_{xx}(0) = \lim_{T\to\infty} \int_{-T}^{T} x^2(t)\,\mathrm{d}t = \overline{x^2(t)}. \tag{4.54}$$

The cross-correlation function is neither even nor symmetric, but

$$\varphi_{xy}(\tau) = \varphi_{yx}(-\tau). \tag{4.55}$$

From Eq. (4.52), substituting $\tau=0$, we get

$$\varphi_{xy}(0) = \varphi_{yx}(0) = \overline{x(t)y(t)}. \tag{4.56}$$

If two random functions – two random signals – are statistically uncorrelated, then the expected product of their values will be equal to the product of their expected values. On this basis it can be proved that statistically independent signals are uncorrelated, i.e. their cross-correlation function is equal to zero.

5. COMPUTING TECHNIQUES

Measurement evaluation may sometimes involve substantial computation, in which an electronic digital computer may be of considerable assistance.

Reference [1] deals with measurement evaluation by an electronic computer. It goes into great detail on the description and application of programs suitable for the evaluation of single measurements, measurement series, point-like and function-like measured data, etc.

The title of this chapter may be misleading because by "computing technique" the actual utilization of electronic digital computers in some defined field, is usually understood. Instead of going into detail, we draw the reader's attention to the book referred to, and in particular to the programs it contains and their uses.

The aim of this chapter is to offer methods which facilitate the evaluation of measured data. These include the now common pocket or desk-type calculator. The chapter should also prepare the reader for the use of an electronic computer.

In what follows, therefore, we shall deal with measured data, their errors, their practical application, logarithmic units and their uses and, generally, with the methods which reduce the work of calculation and have made the formerly ubiquitous slide rule obsolete.

5.1 APPROXIMATE VALUES AND THEIR ERRORS; OPERATIONS WITH APPROXIMATE VALUES

5.1.1 BASIC CONCEPTS

Since the actual value of physical quantities cannot be ascertained by measurements, the deviation between the measured value and the actual, that is correct value (cf. Section 3.1.1), i.e. the error in the measured value, must be estimated. Information (value, sets of values, signals) obtained by measurement always contains errors.

We do not deal here with systematic errors because they can be corrected in various ways (by tables, etc.).

To solve technical tasks related to measurements or measured data, series of mathematical operations consisting of a finite number of steps must be performed. The tasks consist of three clearly distinct parts:

— the establishment of the required mathematical operations i.e. derivation of formula(e), their determination;
— calculations on the basis of defined formulae;
— the estimation of the errors in the measurement result on the basis of the estimated errors in the measured data.

Here is an example of the handling of data containing error(s).

Example

The task is to determine the mass per metre length of a steel rod having a circle cross section. The mass of the cylinder is

$$M = \frac{d\pi}{4} l\varrho.$$

Knowing the values of d, l and ϱ (with tolerances), how can we determine the value of M and how can we estimate its accuracy? Remember that in practical calculations the accuracy of the result is also influenced by the fact that the accurate value of π is normally considered as its approximate value of 3.14.

The essence of the problem i.e. the method of determining, on the basis of the data plus their tolerances, the resultant errors in the result of a given calculation will be obvious from the presentation.

Let us see what error will occur in the results, if typical mathematical operations are performed on quantities containing errors.

5.1.2 APPROXIMATE QUANTITY, APPROXIMATE VALUE, ERROR LIMIT

Since the accurate value of a physical quantity cannot be ascertained, it is replaced by its approximate value (e.g. its approximate value derived by measurement), and its error limit. Physical quantities expressed as their approximate values and error limits are called approximate (physical) quantities.

If the approximate values of physical quantities are denoted by $A, B, C, \ldots$ and their absolute error limits are a, b, c then such quantities can be approximated by $A \pm a, B \pm b, C \pm c, \ldots,$ which are called approximate quantities.

Assuming that the reader is familiar with the basic concept of error, we call his attention to Section 6.2.1. The error limit is a positive value which is less than the absolute value of the error. Both the absolute and relative error limits may be defined. Their relationship is the same as that between the absolute and relative errors.

In some cases, instead of the absolute error limits, it is preferable to determine the relative error limits $\alpha, \beta, \gamma, \ldots.$ of the measured quantities. The relationships between the absolute and the relative error limits are defined by the formulae

$$A \pm a = A\left[1 \pm \frac{a}{A}\right] = A(1 \pm \alpha), \tag{5.1a}$$

$$B \pm b = B\left[1 \pm \frac{b}{B}\right] = B(1 \pm \beta), \tag{5.1b}$$

$$C \pm c = C\left[1 \pm \frac{c}{C}\right] = C(1 \pm \gamma). \tag{5.1c}$$

The relative errors, obtained from a comparison of the sides of the above formulae will be

$$\pm\alpha = \pm\frac{a}{A}, \tag{5.2a}$$

$$\pm\beta = \pm\frac{b}{B}, \tag{5.2b}$$

$$\pm\gamma = \pm\frac{c}{C}. \tag{5.2c}$$

5.1.3 RULES OF APPROXIMATION; ARITHMETIC OPERATIONS WITH APPROXIMATE QUANTITIES

We now examine expressions of the type $(1 \pm \alpha)$ derived from the approximate quantities of the $(A \pm a)$ type. As was seen before, α is assumed to be very small, i.e. $\alpha \ll 1$, but its effect is nevertheless not negligible.

If α, β, γ, ... are small quantities of the first order, then the products $\alpha\beta$, $\alpha\gamma$, $\beta\gamma$, ... are of the second order, and the products $\alpha\beta\gamma$, $\alpha\beta^2$, $\alpha\gamma^2$, $\beta\alpha^2$, $\gamma\alpha^2$, ... etc. are of the third order in size.

Similar rules apply to quotients, i.e. $\alpha\beta/\gamma$, $\alpha\gamma/\beta$, $\alpha^2\beta/\gamma^2$, ... etc. are all very small quantities, of first order.

The fact that α is of first-order smallness, i.e. $\alpha \ll 1$ is shown in the first place by the fact that beside 1 it is additively not negligible, whereas the higher-order small quantities may be negligible, even additively. This allows the formulation of rules for approximate calculations with very small quantities.

For example, take the term $(1 \pm \alpha^2)$ and square it:

$$(1 \pm \alpha)^2 = 1 \pm 2\alpha + \alpha^2 \cong 1 \pm 2\alpha$$

since α^2 is of second-order smallness, it may be neglected. Similarly

$$(1 \pm \alpha)^3 = 1 \pm 3\alpha \pm 3\alpha^2 \pm \alpha^3 \cong 1 \pm 3\alpha.$$

In general, using the binomial theorem,

$$(1\pm\alpha)^n \cong 1\pm n\alpha.$$

Through the concept of raising to powers (fractional, negative etc.) this last formula can be further generalised. We shall return to this later.

Let us concentrate on approximate quantities of the type $(A\pm a)$. Their significance is shown by the fact that essentially every measurement (all one-dimensional measurements, individual points in every continuous recording, single data originating from the evaluation of measurement series – cf. Section 6.2.1 – ect.) yield such expressions.

In the following, we shall examine in detail arithmetic operations on approximate quantities, and the effect on the resultant accuracy of the results. We shall rely upon the rules of approximation already outlined which, by allowing simplifications, facilitate the computing work.

The sum of approximate quantities

Let us write down the sum of an arbitrary number of approximate quantities. Obviously,

$$(A\pm a)+(B\pm b)+(C\pm C)+\ldots = (A+B+C+\ldots)+(\pm a\pm b\pm c\pm\ldots). \quad (5.3)$$

The second parenthesis on the right side consists of the absolute error of the sum of the approximate quantities. It is denoted by Δ

$$\Delta = \pm a\pm b\pm c\pm\ldots. \quad (5.4)$$

The absolute error limit Δ_k is the error limit defined as the maximum error which arises in the most unfavourable case. The sum, using (5.4), is

$$\Delta_k = \pm(|a|+|b|+|c|+\ldots). \quad (5.5)$$

In the same way as the relative error is calculated from the absolute error, we may compute the relative error limit of the sum δ_k, which means that the absolute error limit is compared with the sum of the approximate values:

$$\delta_k = \frac{\Delta_k}{A+B+C+\ldots} = \pm\frac{|a|+|b|+|c|+\ldots}{A+B+C+\ldots}. \quad (5.6)$$

The difference between approximate quantities

Computations are carried out with two approximate quantities, the minuend and the subtrahend. As in the case of the sum (bearing in mind that the sign changes):

$$(A\pm a)-(B\pm b) = A-B+(\pm a\mp b). \quad (5.7)$$

The absolute error limit for the difference is

$$\Delta_k = \pm(|a|+|b|). \quad (5.8)$$

its relative error:

$$\delta = \frac{\pm a \mp b}{A-B}, \tag{5.9}$$

and the relative error limit:

$$\delta_k = \pm\frac{|a|+|b|}{A-B}. \tag{5.10}$$

The product of approximate quantities

As outlined before, but now in detail

$$(A\pm a)(B\pm b)(C\pm c)... = A(1\pm\alpha)B(1\pm\beta)C(1\pm\gamma)... =$$
$$= ABC...(1\pm\alpha\pm\beta\pm\gamma\pm\alpha\beta\pm\beta\gamma\pm\alpha\gamma\pm\alpha\beta\gamma\pm...). \tag{5.11}$$

Neglecting the small quantities of the second and third orders which in (5.11) stand beside 1, we may write down the approximate relation:

$$(A\pm a)(B\pm b)(C\pm c)... = ABC...(1\pm\alpha\pm\beta\pm\gamma\pm...) \tag{5.12}$$

from which the relative error in the product is

$$\delta = \pm\alpha\pm\beta\pm\gamma\pm..., \tag{5.13}$$

and the relative error limit is

$$\delta_k = \pm(|\alpha|+|\beta|+|\gamma|+...). \tag{5.14}$$

The absolute errors can be computed in the same simple way. Neglecting the details of computation, the absolute error is

$$\Delta = \pm\alpha A\pm\beta B\pm\gamma C\pm.... \tag{5.15}$$

and on this basis the absolute error limit is

$$\Delta_k = \pm(|\alpha A|+|\beta B|+|\gamma C|+...). \tag{5.16}$$

The quotient of approximate quantities.

For the sake of simplicity we deal with only two approximate quantities:

$$\frac{A\pm a}{B\pm b} = \frac{A}{B}\frac{1\pm\alpha}{1\pm\beta}. \tag{5.17}$$

The next step after Eq. (5.17) is to divide the polynomials $(1\pm\alpha):(1\pm\beta)$. As in the considerations of the criteria of approximation:

$$(1\pm\alpha):(1\pm\beta) = 1-\frac{\alpha\pm\beta}{1\pm\beta} \cong 1\pm\alpha\mp\beta, \tag{5.18}$$

and the relative error

$$\delta = \pm\alpha\mp\beta, \tag{5.19}$$

and the relative error limit

$$\delta_k = \pm(|\alpha| + |\beta|). \tag{5.20}$$

The limit of the absolute error in the quotient can be determined by multiplying Eq. (5.20) by the quotient A/B

$$\Delta_k = \pm(|\alpha| + |\beta|)\frac{A}{B} = \pm\left(\left|\frac{a}{A}\right| + \left|\frac{b}{B}\right|\right)\frac{A}{B} = \pm\frac{|a|B + |b|A}{B^2}. \tag{5.21}$$

The absolute error in the quotient is

$$\Delta = \pm\frac{aB \pm bA}{B^2}. \tag{5.22}$$

Raising approximate quantities to powers

In the previous section we touched upon the method of raising approximate quantities to powers. For an arbitrary approximate value:

$$(A \pm a)^n = A^n(1 \pm \alpha)^n = \left(1 \pm n\alpha + \frac{n(n-1)}{1 \cdot 2}\alpha^2 \pm \frac{n(n-1)(n-2)}{1 \cdot 2 \cdot 3}\alpha^3 + \ldots\right). \tag{5.23}$$

Neglecting the small quantities of the second and higher orders

$$(A \pm a)^n = A^n(1 \pm \alpha)^n \cong A^n(1 \pm n\alpha). \tag{5.24}$$

From this, the relative error is

$$\delta = \pm n\alpha, \tag{5.25}$$

and the absolute error

$$\Delta = \pm A^n n\alpha = \pm A^n n\frac{a}{A} = -nA^{n-1}a. \tag{5.26}$$

Since partial errors do not appear additively in the end result, they also indicate the error limits.

Extraction of roots from approximate quantities

Since the extraction of roots can be treated as raising to a fractional power, the above relationship can be generalised for the extraction of roots:

$$\sqrt[n]{A \pm a} = \sqrt[n]{A}\sqrt[n]{1 \pm \alpha}. \tag{5.27}$$

According to (5.25) the relative error in a root is

$$\delta = \pm\frac{1}{n}\alpha, \tag{5.28}$$

and the absolute error

$$\Delta = \pm\frac{1}{n}\frac{a}{A} = \pm\frac{1}{n}A^{(1/n)-1}a. \tag{5.29}$$

Up to now we have dealt with the more important rules relating to the basic operations, and the resultant errors involved in the basic operations with approximate quantities. Now we shall present two examples, dealing with the most common cases of basic operations. The first is the calculation of a rational fraction for approximate quantities, and the second is that of the reciprocal of an approximate quantity.

Rational fraction from approximate quantities

$$\frac{(A\pm a)(B\pm b)\ldots}{(C\pm c)(D\pm d)\ldots}=\frac{AB\ldots}{CD\ldots}\cdot\frac{(1\pm\alpha)(1\pm\beta)\ldots}{(1\pm\gamma)(1\pm\delta)\ldots}. \tag{5.30}$$

The absolute error is

$$\Delta=\pm a\pm b\pm\ldots\mp c\mp d\mp\ldots. \tag{5.31}$$

and the relative error δ' (the comma is used to distinguish between the two δ-s):

$$\delta'=\pm\alpha\pm\beta\pm\ldots\mp\gamma\mp\delta\mp\ldots. \tag{5.32}$$

The absolute error limit is

$$\Delta_k=\pm(|a|+|b|+\ldots+|c|+|d|+\ldots). \tag{5.33}$$

The reciprocal of approximate quantities

Starting from the general term, let us calculate the reciprocal:

$$\frac{1}{A\pm a}=\frac{1}{A}(1\mp\alpha). \tag{5.34}$$

The absolute error is

$$\Delta=\mp\alpha, \tag{5.35}$$

and the relative error is

$$\delta=\mp\alpha A. \tag{5.36}$$

The rules relating to the basic operations and the errors involved in them may also be applied to more complicated relationships and algorithms. Although, as stated before, the approximate quantities obtained by measurement have the form $(A\pm a)$, for practical computation the form $(\pm\alpha)$ is perferable. The approximate formulae for the case $\alpha\ll 1$ have been compiled in Table 5.1, together with the approximate formulae derived by raising to powers.

In practical approximate computation, approximate formulae used instead of certain (mostly transcendent) functions assume considerable importance. In exact computation, these functions are obtained from tables of functions calculated from their infinite series.

A single or a few members of these series can be used in approximate formulae. In Table 5.2 the reader will find a number of frequently-used approximate formulae together with relevant information and/or remarks.

To sum up approximate formulae offer a great advantage. For frequently-used algorithms (e.g. non-linear combinations of $(1\pm\alpha)$, its powers, some transcendent

TABLE 5.1.

Approximate formulae using $(1\pm\alpha)$ *in the case* $\alpha\ll 1$

Number	The term	The approximate formula	Remarks
1	$(1\pm\alpha)^n$	$1\pm n\alpha$	
2	$\frac{1}{1\pm\alpha}$	$1\mp\alpha$	
3	$\sqrt{1\pm\alpha}$	$1\pm\frac{\alpha}{2}$	
4	$\frac{1}{\sqrt{1\pm\alpha}}$	$1\pm\frac{\alpha}{2}$	
5	$\sqrt[n]{1\pm n\alpha}$	$1\pm\alpha$	
6	$(1\pm\alpha)(1\pm\beta)\ldots$	$1\pm\alpha\pm\beta\pm\ldots$	
7	$\frac{(1\pm\alpha)(1\pm\beta)\ldots}{(1\pm\gamma)(1\pm\delta)\ldots}$	$1\pm\alpha\pm\beta\pm\ldots\mp\gamma\mp\delta\mp\ldots$	
8	$(A\pm a)^n=[A(1\pm\alpha)]^n$	$A^n(1\pm n\alpha)$	The generalisation of 1 with $\alpha=1/A$ (for the derivation see Section 5.1.4)

functions etc.) for low values of the independent variable, they substitute simple operations for the original formulae or parts of the formulae, and give simpler numerical computations.

Their shortcoming lies in the fact that a certain amount of experience is needed to identify cases where the formulae can be applied without serious detriment to accuracy.

However, the method can easily be acquired by solving a number of problems.

5.2 LOGARITHMIC UNITS, LEVELS

In science and technology when – as often happens – computed or measured values vary over wide ranges; co-ordinate axes and functions are scaled in so-called logarithmic units, for reasons the most important of which are:

— wide limits embrace several orders of magnitude which cannot conveniently be covered by linear scales;

TABLE 5.2.

Approximate formulae of transcendent functions in the case of low argument values

The function	The approximate formula	Remarks
e^{α}	$1+\alpha$	Better approximation: $1+\alpha+\frac{\alpha^2}{2}$
a^{α}	$1+\alpha \ln a$	
$\ln \alpha$	$\alpha-\frac{\alpha^2}{2}$	Valid provided that $\|\alpha\| \ll 1$ and $\alpha \neq 0$
$\sin \alpha$	α	Better approximation: $\alpha-\frac{\alpha^3}{6}$
$\cos \alpha$	1	Better approximation: $1-\frac{\alpha^2}{2}$
$\tan \alpha$	α	Better approximation: $\alpha+\frac{\alpha^3}{3}$
$\text{cotan}\ \alpha$	$\frac{1}{\alpha}$	Better approximation: $\frac{1}{\alpha}-\frac{\alpha}{3}$; $(\alpha \neq 0)$
$\text{sh}\ \alpha$	α	Better approximation: $\alpha+\frac{\alpha^3}{6}$
$\text{ch}\ \alpha$	1	Better approximation: $1+\frac{\alpha^2}{2}$
$\text{th}\ \alpha$	α	Better approximation: $\alpha-\frac{\alpha^3}{3}$
$\text{cth}\ \alpha$	$\frac{1}{\alpha}$	Better approximation: $\frac{1}{\alpha}+\frac{\alpha}{3}$; $(\alpha \neq 0)$
$\sin(\varphi \pm \alpha)$	$\sin \varphi \pm \alpha \cos \varphi$	$\alpha \ll 1$, φ arbitrary

N. B. We draw the reader's attention to the fact that, since the approximate formulae originate from the series for transcendent functions, the values of the independent variables should be expressed in radians $\left(\widehat{\alpha}=\alpha^{\circ}\frac{\pi}{180^{\circ}}\right)$.

— the logarithm simplifies certain mathematical operations and certain types of function. In measurement evaluation certain graphical methods have particular significance, and in this field the logarithmic co-ordinate system can be applied to advantage;

— a large number of natural phenomena are essentially logarithmic in character. The sensitivity of the human ear, for example, is logarithmic and therefore acoustics uses logarithmic units almost exclusively.

Loudness is a subjective characteristic of human hearing; its relevant objective quantity is volume (or intensity) of sound. The relationship between the subjective loudness and the objective sound intensity is a function of frequency and amplitude and this relationship is logarithmic.

Before going into more detail on logarithmic units, let us remember some important facts about logarithms. We shall deal with two types of logarithm, $\log x$ and natural with base $e=2.718$.

The definition of logarithm leads to the following fundamentally important identity:

$$10^{\log x} = e^{\ln x} = \log 10^x = \ln e^x = x. \tag{5.37}$$

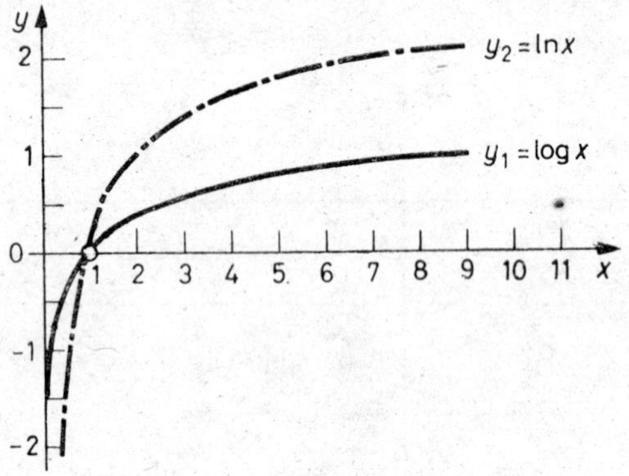

Fig. 5.1. *The curves for* log *x and* ln *x*

The two logarithmic values of the functions are shown in Fig. 5.1. As can be seen, in the range of positive values of the functions (argument values above 1) the natural logarithm exceeds logarithm to base 10. Their relationships can be derived from Eq. (5.37), namely

$$\log x = \ln x \log e. \tag{5.38}$$

The logarithm to base 10 of the base of natural logarithms is known as the modulus of conversion and is denoted by M:

$$M = \log e = 0.4342. \tag{5.39}$$

To determine $\ln x$ the reciprocal of M is used, in the rearranged form of (5.39), with $1/M=2.3$.

The logarithm of the complex number

$$\ln z = \ln |z| + j \operatorname{arc}(z), \tag{5.40}$$

yields an important relationship.

The logarithmic method was first used in electrical systems, for the calculation of power ratio and power transmission. For transmission engineering, Graham Bell recommended the logarithmic method to calculate the ratio of power measured at the input and output of a system.

Before going into details, we shall make a detour, but with the limitations in mind, briefly, and only to prove that the method accepted and now classic in electricity can be applied to other systems (mechanical, flow, acoustic, thermal, etc.) as well. Such generalisation is possible in the so-called general systems theory and, although we do not wish to go into great detail, we shall try to present the more important elements of the theory and call the reader's attention to the relevant literature [5.1], [5.2], [5.3], [5.4].

Systems can be described by so-called co-ordinates of state which may be intensive (network co-ordinates) or extensive (quantitative co-ordinates.) Table 5.3 illustrates intensive and extensive state co-ordinates of different systems.

The table shows that systems using dissimilar energies can be described in very similar ways. For this book we shall need only the intensive (network) co-ordinates which define the concepts of impedance and power, shown in the last two columns. In addition to the known systems, the last row of the table describes an imaginary — abstract linear — system. It is used for convenience because it enables conclusions to be drawn about relationships and extended to all kinds of systems.

As always, this abstract system too has voltage- and current-type intensive co-ordinates whose quotient has the character of impedance, and whose product yields the power.

Now, reverting to logarithmic units, this abstract system will be used in writing down the relationships to indicate the fact that they hold true whatever the quantities concerned, be they mechanical, electrical, acoustical, etc.

In the logarithmic method suggested by Bell for transmission engineering, the logarithm with base 10 is called bel (B) in the memory of Bell

$$a^{B} = \log\frac{P_1}{P_0}\ \mathrm{B}, \tag{5.41}$$

where P_1 denotes the power and P_0 the reference power.

Both powers should be described in the same units.

In practice a tenth of a bel, the decibel (dB) is used:

$$a^{\mathrm{dB}} = 10\log\frac{P_1}{P_0}\ \mathrm{dB}. \tag{5.42}$$

The other relative logarithmic unit is the neper (Np). It is the natural logarithm of the quotient of voltage-type quantities (e.g. voltages, pressures, forces, sound pres-

TABLE 5.3.

Extensive and intensive state co-ordinates. Power calculation

System	State co-ordinate				Generalised impedance	Power
	Intensive (network)		Extensive (quantitative)			
Electrical	Voltage $U = \frac{d\Phi}{dt}$	Current $I = \frac{dQ}{dt}$	Flux $\Phi = \int U\,dt$ (voltage surge)	Charge $Q = \int I\,dt$ (current surge)	$Z = \frac{U}{I}$	$P = UI = \frac{U^2}{Z} = I^2 Z$
Mechanical: translatory	Force $F = \frac{dI}{dt}$	Speed velocity $v = \frac{dx}{dt}$	Pulse $I = \int F\,dt$	Shift $x = \int v\,dt$	$Z_{m,trans} = \frac{F}{v}$	$P = Fv = \frac{F^2}{Z_{m,trans}} = v^2 Z_{m,trans}$
Mechanical: rotary	Torque $M = \frac{d\pi}{dt}$	Angular velocity $\omega = \frac{d\varphi}{dt}$	Spin $\pi = \int M\,dt$	Angular rotation $\varphi = \int \omega\,dt$	$Z_{m,rot} = \frac{M}{\omega}$	$P = M\omega = \frac{M^2}{Z_{m,rot}} = \omega^2 Z_{m,rot}$
Thermal	Temperature ϑ	Entropy flow $s = \frac{dS}{dt}$		Entropy flow $S = \int s\,dt$	$Z_t = \frac{\vartheta}{s}$	$P = \vartheta = \frac{\vartheta^2}{Z_t} = s^2 Z_t$
Flow (and acoustic)	Pressure $p = \frac{d\gamma}{dt}$	Volumetric flow $q = \frac{dV}{dt}$	Pulse per unit surface area $\gamma = \int p\,dt$	Volume $V = \int q\,dt$	$Z_{gen} = \frac{p}{q}$	$P = pq = \frac{p^2}{Z_{gen}} = q^2 Z_{gen}$
Generalised (abstract linear) system	Generalised voltage $\mu = \frac{d}{dt}$	Generalised current $\lambda = \frac{d}{dt}$	Generalised voltage surge $\int \mu\,dt$	Generalised current surge $\int \lambda\,dt$	$Z_{gen} = \frac{\mu}{\lambda}$	$P = \mu\lambda = \frac{\mu^2}{Z_{gen}} = \lambda^2 Z_{gen}$

sures, etc.). With general voltages:

$$a^{\mathrm{Np}} = \ln \frac{\mu_1}{\mu_0} \mathrm{Np}, \tag{5.43}$$

where μ_1 is the general voltage concerned; and μ_0 the reference value.

Since in linear systems or in systems assumed to be linear the potentials and powers are in unambiguous relationships, through impedance, as seen in Table 5.3; knowing the impedances we can express the decibel by voltages or neper by powers. Let us assume that the general impedance Z_0 corresponds to the power P_0 and the general impedance Z_1 corresponds to P_1. In this case the decibel value calculated from the voltages is as follows:

$$a^{\mathrm{dB}} = 10 \log \frac{P_1}{P_0} = 10 \log \frac{\mu_1^2 Z_0}{\mu_0^2 Z_1} = 20 \log \frac{\mu_1}{\mu_0} + 10 \log \frac{Z_0}{Z_1}. \tag{5.44}$$

If the two impedances are equal, then $\log \frac{Z_0}{Z_1} = \log 1 = 0$ and so

$$a^{\mathrm{dB}} = 20 \log \frac{\mu_1}{\mu_0}. \tag{5.45}$$

Naturally, the neper too can be expressed in terms of power:

$$a^{\mathrm{Np}} = \ln \frac{\mu_1}{\mu_0} = \ln \sqrt{\frac{P_1 Z_1}{P_0 Z_0}} = \frac{1}{2} \ln \frac{P_1}{P_0} + \frac{1}{2} \ln \frac{Z_1}{Z_0}. \tag{5.46}$$

If the two impedances are again equal, then because $\ln \frac{Z_1}{Z_0} = \ln 1 = 0$ we will have

$$a^{\mathrm{Np}} = \frac{1}{2} \frac{P_1}{P_0}. \tag{5.47}$$

It is worth noting that (5.45) and (5.47) apply even if the criterion $Z_0 = Z_1$ is not met. However, if (5.44) and (5.46) are taken as the definitions of dB and Np respectively, then without knowing the impedance ratios no statements can be made with any certainty regarding the power and voltage ratios.

If the reference values (power, generalised voltage) of a system (machine, electric circuit, interconnected equipment, etc.) are identical throughout, then the values of dB and neper measured at different points are called the levels (power level, voltage level).

Given the impedance ratio, there is a conversion between the decibel and the neper which can be expressed using the modulus of conversion between the two different logarithms (cf. (5.39)). Conversion is generally achieved if $Z_1 = Z_0$. Making use of (5.39) and (5.45), we have

$$a^{\mathrm{dB}} = 20 \log \frac{\mu_1}{\mu_0} = 20 \times 0.4342 \ln \frac{\mu_1}{\mu_0} \cong 8.7 a^{\mathrm{Np}}. \tag{5.48}$$

Accordingly, the

$$a^{dB} = 8.7a^{Np} \tag{5.49a}$$

and

$$a^{Np} = 0.115a^{dB} \tag{5.49b}$$

conversion equations hold true. From this it follows that 1 Np=8.7 dB and that 1 dB=0.115 Np. Equations (5.49) are valid even if the dB is defined by Eq. (5.45), but the powers in that case can only be defined if the impedances are known.

TABLE 5.4.

Decimal ratios and the corresponding logarithmic levels

Decimal ratio $\frac{x}{x_0}\left(\frac{\mu_1}{\mu_0}\right)$	$a^{dB} = 20\log\frac{\mu_1}{\mu_0}$ dB	$a^{Np} = \ln\frac{\mu_1}{\mu_0}$ Np
0.001	−60	−6.9
0.01	−40	−4.6
0.1	−20	−2.3
0.2	≈ −14	−1.6
0.5	≈ −6	−0.693
0.9	≈ −0.92	−0.105
1.0	0	0
1.1	≈ +0.83	+0.0953
1.2	≈ −1.6	+0.182
2	≈ +3	+0.346
2	≈ +6	+0.693
3.16	+10	+1.15
10	+20	+2.3
20	≈ +26	+2.995
100	+40	+4.605
1000	+60	+6.907

In Table 5.4 the more frequent decimal ratios and logarithmic levels have been compiled in terms of dB and Np, with reference to the ratios of generalised voltages.

As regards logarithmic units, it should be remembered that while the neper is used mainly in telecommunication, the decibel was accepted in practically all fields of technology (automatic control, oscillations, acoustics, turbulence, etc.) Both may be used to advantage, and the choice depends on the type of problem in hand. We again draw the reader's attention to impedances and their relationships. This is important, and not only in electrical problems. With the spread of the systems theory concept, the mechanical, acoustical, hydraulic, pneumatic, etc. impedances have found extensive applications.

The logarithmic units decibel and neper provide values relative to the power levels measured in various systems, to generalized intensive co-ordinates (generalised voltage, generalised current) to one another or to other values outside the system. These relative levels and some other characteristic quantities of a system, expressed in some decimal (e.g. SI) units may vary between wide limits, depending on frequency. In such cases, to cover a sufficiently wide range of frequencies, it is advisable to state the frequency too in logarithmic (relative) units. The logarithmic (relative) units of frequency generally applied are the decade, the octave (1/2 octave) and the 1/3 octave, also called "third". Relative frequencies expressed in logarithmic units are denoted by ν and, as in the case of the decibel and the neper, the unit concerned is in the superscript: decade=D, octave=oct., 1/3 octave=third.

To calculate the relative units of frequency the logarithm of the frequency quotient on the basis of the required frequency coverage, is considered.

Decade

The decade is a frequency coverage often, $f_2/f_1=10$, therefore

$$\nu^{\mathrm{D}} = \log \frac{f_2}{f_1}. \tag{5.50}$$

Octave

The octave is a frequency overlapping of 2, $f_2/f_1=2$. Accordingly,

$$\nu^{\mathrm{oct}} = {}^{2}\log \frac{f_2}{f_1}. \tag{5.51}$$

Third

The third (called also 1/3 octave) means $\sqrt[3]{2}$ a frequency overlapping of 3 i.e. $f_2/f_1=\sqrt[3]{2}$. Thus,

$$\nu^{\mathrm{third}} = {}^{\sqrt[3]{2}}\log \frac{f_2}{f_1}. \tag{5.52}$$

As in (5.37),

$$10^{\log x} = 2^{{}^{2}\log x} = (\sqrt[3]{2})^{{}^{\sqrt[3]{2}}\log x} \tag{5.53}$$

or taking logarithms to base 10,

$$\log x = {}^{2}\log x \log 2 = {}^{\sqrt[3]{2}}\log x \log \sqrt[2]{2}. \tag{5.54}$$

Since $\log_{10}=0.30103$; $1/\log_{10}=3.33$

$$\log_{10} \sqrt[3]{2} = 0.1003433; \; 1/\log_{10} \sqrt[3]{2} = 9.96 \cong 10$$

between the frequencies given in different logarithmic (relative) units the following relationships exist:

$$\begin{aligned} 1 \text{ decade} &= 3.33 \text{ octave (a)} \\ 1 \text{ decade} &= 10 \text{ thirds (b)} \\ 1 \text{ octave} &= 0.3 \text{ decade (c)} \\ 1 \text{ octave} &= 3 \text{ thirds (d)} \\ 1 \text{ third} &= 0.1 \text{ decade (e)} \\ 1 \text{ third} &= 0.33 \text{ octave (f)} \end{aligned} \qquad (5.55a\text{–}f)$$

In accordance with the conversion formulae (5.55), to facilitate the conversion from logarithmic to decimal units of frequency, double-scaled axes can be produced (Fig. 5.2).

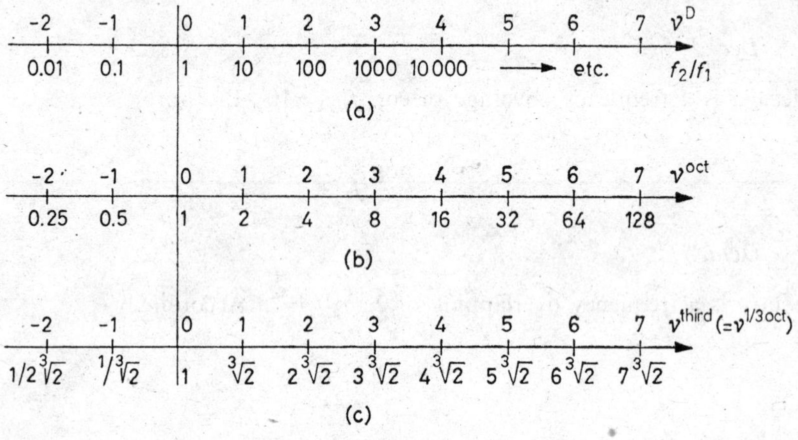

Fig. 5.2. *Logarithmic–decimal double frequency axes:* (a) *decade scale,* (b) *octave scale,* (c) *third scale*

5.3 AIDS USED IN THE NUMERICAL EVALUATION OF MEASUREMENTS

The evaluation of measurement results may involve a considerable amount of numerical computation. As mentioned, for this an electronic digital computer may be used to advantage. Reference [1] deals in detail with its use. In this chapter we shall restrict ourselves to surveying the manual aids and their uses. By "manual aids" we mean the slide rule and the electronic calculator, i.e. electronic pocket or desk-type computer.

The slide rule is the classic type of aid for technical calculation, and its manipulation is widely known. Although its importance has diminished with the spread of electronic calculators, in certain cases, e.g. when an electronic calculator is not accessible or available, it is still used. Its shortcoming is its limited accuracy, so we shall confine our treatment to pointing out the possibilities of increasing its accuracy.

The calculator is gradually assuming the role of an "electronic slide rule". In what follows we shall try to describe its uses and its potentials.

5.3.1 THE USE OF THE SLIDE RULE; WAYS TO INCREASE ITS ACCURACY

In the technical calculations of our day the slide rule can no longer cope. This section has been inserted for historical reasons.

We assume that the reader is familiar with the slide rule and its use.

The slide rule allows the execution of all basic operations except addition and subtraction. Its accuracy increases in proportion to its length. The most frequently used sizes are as follows:

- pocket-type slide rule, 12.5 cm,
- manual slide rule, 25 cm,
- drafting room slide rule, 50 cm.

The slide rules in widest use are 25 cm long and are suitable for multiplication and division with an error of approximately 0.1%. This is equal to the accuracy of calculations on paper, with two digits. The smaller (12.5 cm) and longer (50 cm) types work with an accuracy of one digit less or more, respectively.

The accuracy of the slide rule can be increased in two ways:

- by doing calculations on paper with the aid of a slide rule, the slide rule serving as an intermediate aid (e.g. in division to indicate how many times the divisor is contained in the dividend, etc.)
- by the application of special techniques.

To illustrate the above, we quote two cases which may increase the accuracy of the slide rule in the two most frequent arithmetical operations, i.e. multiplication and division.

Multiplication

A product ab can be expressed in the form:

$$ab = a(c+d), \tag{5.56}$$

where, naturally,

$$c+d = b. \tag{5.57}$$

If the value of c is a power of 10 then ac can be determined by the proper transposition of the decimal point, i.e. essentially, by a mental calculation. In this case it is only the $|d|$ $(d=b-c)$ multiplicand that must be multiplied by a with the aid of the slide rule. The result will be more accurate than that obtained by the determination of the value of ab directly with the slide rule, particularly if $|d| \ll c$.

Example

Let two numbers to be multiplied be $a=1.26$ and $b=20.981$. Performing the operation directly with a 25 cm slide rule we have

$$ab = 1.26 \times 20.981 = 26.4.$$

Applying the method outlined above and choosing the values $c=2\times 10$ and $d=$ $=0.981$, $c+d$ will give $20+0.981=20.981=b$ and the product assumes the form of

$$ab = a(c+d) = 1.26(2\times 10+0.981).$$

The first multiplication may be done mentally: $1.26\times 0.981=1.236$. Adding the two members:

$$25.2+1.236 = 26.436$$

gives a more accurate result.

Division

The quotient a/b may appear in the following form

$$\frac{a}{b} = \frac{b+d}{b} = 1+\frac{d}{b}, \tag{5.58}$$

where $d=a-b$ with d being positive or negative. The quotient d/b can be computed with a slide rule and the result will be more accurate than that obtained by using the slide rule directly, especially if $|d| \ll b$.

Example

Let

$$a = 56.88 \quad \text{and} \quad b = 46.43.$$

With the direct use of the slide rule we shall have

$$\frac{a}{b} = \frac{56.88}{46.43} = 1.23$$

but a calculation with $d=(a-b)=56.88-46.43=10.45$ we get a more accurate result, i.e.

$$\frac{a}{b} = \frac{b+d}{b} = 1+\frac{10.45}{46.43} = 1+0.225 = 1.225$$

Difference between quotients

When computing the value of $[(a/b)-(c/d)]$, serious errors may arise if the values of the numerator and denominator of each quotient are close to one another. In such cases, the computation should take place on the basis of the following identity:

$$\frac{a}{b}-\frac{c}{d}=\frac{d-c}{d}-\frac{b-a}{b}. \tag{5.59}$$

Example

$a=44$, $b=45$, $c=23$, $d=24$ are given:

$$\frac{44}{45}-\frac{23}{24}=\frac{24-23}{24}-\frac{45-44}{45}=\frac{1}{24}-\frac{1}{45}=0.042-0.0222=0.0188.$$

5.3.2 ELECTRONIC CALCULATORS

Electronic calculators are the most frequently used form of microcomputers. Since the members of their family are rather heterogeneous, it is difficult to deal with all of them in the same way [5.5], [5.6], [5.8], [5.9], [5.10]. In what follows we shall call all those electronic devices which can execute the four basic operations "calculators", be they simple pocket calculators, more sophisticated types that can produce functions and have storage media added, or programmable pocket or desk-top calculators capable of producing many functions and executing a variety of operations. They facilitate general tasks and, in particular, help people engaged in measurement evaluation.

The above means that the term "calculator" will be used to refer to a far from uniform family of products. Since the types in widest use in engineering calculations are the calculators of medium complexity, we shall concentrate on these. When we deal with other types, we shall say so.

In this short survey we cannot go beyond the basic and most important features of calculators. Interested readers should study References [5.6], [5.7], [5.8], [5.9] and [5.10], and manuals on calculators and instructions for their use.

The logic structure of calculators

Calculators have a basic logic structure which may be

— arithmetic logic,
— algebraic logic or
— so-called inverse Polish notation.

If the calculator uses arithmetic logic, it has no separate key $\boxed{=}$. The $=$ sign is inputted using the same key as $+$ and $-$ $\boxed{+=}$ and $\boxed{-=}$. With such calculators, first the digit and then the symbol for the operation is fed in, e.g.

$$\boxed{6}\,\boxed{+=}\,\boxed{2}\,\boxed{-=}$$

and in the display the result 4 will appear.

Calculators with algebraic logic have a separate $\boxed{=}$ key. Their most important characteristic is that the sequence in which the digits and operations are fed in, is the same as in the spoken language, e.g.

$$\boxed{6}\,\boxed{-}\,\boxed{2}\,\boxed{=}$$

i.e. "six minus two is equal to"..., again with 4 appearing in the display. In the family of calculators with algebraic logic, two more subvariants, in addition to the above basic form, exist:

— calculators with algebraic logic and hierarchy,
— calculators with algebraic logic and keys for brackets.

In calculators with hierarchy the operations of "higher order", called point operations (multiplication, division) have priority over the operations of "lower order", known as line operations (addition and subtraction).

With such a calculator the result of keying in

$$\boxed{2}\,\boxed{\times}\,\boxed{2}\,\boxed{+}\,\boxed{3}\,\boxed{\times}\,\boxed{4}$$

would be 16. Due to hierarchy, $(2\times2)+(3\times4)=16$, while a calculator without hierarchy would produce the result for $(2\times2+3)\times4=28$.

In calculators with algebraic logic and provided with bracket keys, the grouping of the operations is not determined by the calculator itself (on the basis of hierarchy) but by the operator who uses the bracket keys in the same way as in calculations on paper.

Some calculators can produce several pairs of brackets, but these are seldom used for technical calculations.

The "inverse Polish notation" was introduced after the work of Lukasiewicz, the Polish mathematician. From what went before, it will be clear that the organisation of the successive operations (e.g. use of brackets, sequence of operations, etc.) causes certain problems. Inverse Polish notation solves these and ensures complete unambiguity. Such calculators have no $\boxed{=}$ key for the result but they have one which is found in calculators of other types

$$\boxed{\uparrow} \text{ or } \boxed{\text{ENTER}\uparrow} \text{ or } \boxed{\text{ENTER}}$$

Calculator structure

The structure of a calculator is independent of its basic logic structure. The user sees the following important structural parts:

— casing,
— keyboard,
— display,
— on/off switch,
— printer (in some desk-top types).

In addition to these clearly visible units, the following functional units are inside:

— computing circuit,
— registers,
— memory(ies),
— supply unit, etc.

We shall now select the more important structural and functional units for further discussions:

— keyboard,
— keys serving different purposes,
— "enter" keys,
— operation keys,
— other keys.

In the simplest types of calculator the input keys are the digits 0 to 9 and the decimal point (or comma). The more sophisticated models also have keys for changing the sign [+/−] or [CHS] and some for the most frequently occurring values:

$\pi = 3.141\ 592\ 6\ldots$
$e = 2.718\ 28\ldots$
$g = 9.81\ldots.$

Some calculators have keys for [%] and [Y^x], the latter to produce powers (even fractional ones) on the basis of the relationship $y^x = e^{x \ln y}$.

Further input keys may be [EXP] or [EE] (to enter exponents and to produce numerals with floating point). The latter will be dealt with later.

The operation keys control the execution of mathematical or other operations on the numbers entered into the calculator. Assuming the simpler algebraic structure, the more important operation keys are as follows:

[+] [−] [×] [÷] for the four basic operations

[=] result

[C] rep. [AC] (for clear or all clear).

Note: Some calculators have both [C] and [AC]. AC clears every number previously introduced except those in the memory, while C clears only the number fed in last. Thus, if a number has been erroneously entered, it is not necessary to clear the entire calculation, only the last number. In other calculators the functions of C and AC are united in the same key, whose first depression executes the function of C, the second that of AC.

In more complicated calculators, in addition to the basic operations there are also keys for

— functions,
— storage (memory),
— the organisation of calculations.

The function keys retrieve the value of mathematical and other functions on the basis of the argument values, using the calculator as a function table. In the more important technical calculations various exponential (e^x, x^y), logarithmic (log, ln), trigonometric (sin, cos, tan, cot) and hyperbolic (sh, ch, th, cth) functions as well as their inverses (arcsin, arccos, arctan, arccotan, arsh, arch, arth, arcoth), exponential and radical functions (x^2, x^3, $\sqrt{x}$, $\sqrt[3]{x}$)... may occur.

When the keyboard is too small to accommodate the large number of function keys, the manufacturers assign two or sometimes even three functions to the same key. In such calculators each key has 2 or 3 symbols, and for the retrieval of the functional values an [F] (function) or [2nd] auxiliary key is provided. To retrieve inverse functional values the auxiliary keys [$\sin^{-1}$] [$\tan^{-1}$] etc. or [arc] are used.

Note: In calculators without function keys, the values of the more important functions can be approximated by the first few members of their infinite series.

At least two registers are provided even on the simplest calculators, irrespective of their logic structure. They are denoted X and Y and serve the following purposes:

— to receive the so-called operands of the calculation,
— to link the two operands in the two registers in the way determined by the manipulation of the keys.

The two registers are called:

X - entry and/or display register,
Y - calculator register.

Into the X register are entered all the digits which the operator enters with the appropriate keys, including the + or − sign. These digits – those of the operands – appear in the display. When the proper operation key is pressed, the contents of the X register are copied into register Y, because the 2 operands require 2 registers. Thereafter the second operand may be entered.

When the result key is pressed, the calculator executes the operation previously entered. The result is stored in the X register and appears simultaneously in the display.

On some calculators an [X↔Y] key is also provided.

The data in the two registers (Y and X) are exchanged when it is pressed.

Memory (storage) keys are found on special calculators. For technical calculations it is advisable to use devices with at least one memory. This is important because many calculations yield intermediate results which with a calculator lacking a memory, would have to be recorded on paper and then entered repeatedly. Memories make such extra work unnecessary.

There are

— simple,
— calculating and
— addressable

memories. The keys for simple memories bear the sign [M] (memory). If such a key is pressed the contents of the X register are transferred to the memory after any digits already there have been cleared. The recall of the digits from the memory is achieved by pressing the [MR] or [RM] (memory recall) key which causes the transfer of the digits stored into register X. Accordingly, numbers fed into a simple memory can be stored or recalled, until new numbers are entered or until the contents of the memory are cleared. Clearing takes place upon pressing the [CM] (clear memory) key.

Calculating memories are especially advantageous in technical calculations and in measurement evaluation. They involve the following keys:

[M+] [M−] [M×] [M÷]

Most calculators have the first two keys, the more complex types have all four. When the key is pressed, the calculator executes the operation according to the lettering on the key (adds, subtracts, etc. the contents of the register X). For instance, by depressing the key [M+] the contents of the entry register are added to the contents of the memory.

Note: Some calculators – less frequently used but particularly well suited for measurement evaluation – feature an [M+X²] key as well. This key can produce scatter. Let us take the case when the content of the memory is 28. By depressing the keys [5] and [M+X²] 53 will appear in the display, since $28+5^2=53$.

Addressable memories are provided only on the most sophisticated, mostly programmable calculators. They feature the following keys:

[STO] [*n*] $(n = 0, 1, 2, \ldots, N)$.

The first key (store) stands for the memory, the second for the latter's address. Thus, to transfer the number with the address n into the X register from the memory, the keys |RCL| |n| or |CL| |n| (RCL=recall, CL=clear) must be depressed.

Display

The display forms part of every (pocket or desk-top) calculator and shows the actual operand or the result. Mostly the so-called seven-segment or mosaic displays are used, very often semiconductor LED's (light-emitting diodes) and the liquid crystal.

Display of numerals on a calculator

Numerals on calculators can be represented in

— fixed-point or
— floating-point

mode. The numbers are usually entered in the same way as they are written, including the decimal point. The entry of a digit is accomplished by pressing an operation key or the result key. If the inverse Polish notation is used, it is done by depressing the |ENTER| key.

In fixed-point notation the most important characteristic is the number of digits which can be held in the calculator. This number is usually eight. The numbers can vary between 99999999 and 0.0000001. The purely fixed-point calculators cannot accept under- or overflow (e.g. 0.0000001...0.0000001=Ø).

In floating-point notation the number z is expressed by the relationship $z=m\times 10^e$ where m is the mantissa and e the exponent. Six or eight digits can be used for the mantissa, and on the display the mantissa and the exponent appear side by side. Thus, e.g. the meaning of $-1.234\,567\,8$ 06 is $-1.234\,567\,8\times 10^6=-1\,234\,567.8$ and that of 1.234 567 8−06 is $1.234\,567\,8\times 10^{-6}=0.000\,001\,234\,567\,8$. Numbers so written are also known as normalised numbers. In order to calculate with numbers in such a form, that is, to enter fixed-point numbers in this form, some calculators are fitted with an |EXP| key. To enter the Boltzmann constant, whose value $k=1.380\,66\times 10^{-23}$ J/K calls for the following sequence:

|1| |·| |3| |8| |0| |6| |6| |EXP| |2| |3| |+/−|

Note: Modern calculators work with more digits than their displays can accommodate. This means that although iterative (rounding) errors may occur, they are negligibly small; e.g. an eight-digit calculator produces logarithms with approximately the same accuracy as a seven-digit logarithmic table. Special attention should be paid to rounding errors in operations where the order of magnitude of the operands is widely different, as for instance 12 345 678+0.001 234 56.

So far we have outlined the manipulation of calculators, the possibilities they offer, etc. Now we shall outline a number of techniques which help in their manipulation, with stress on the less obvious aspects. It should be noted that for technical calculations some unusual techniques may be of great significance. But let us note first of all, even though it may sound obvious that for skill in the use of calculators practice overrides every other requirement.

Series of additive operations may take place in the sequence as they are written, e.g.

$$a+b+c-d+e-f-g+-...$$

may take place as it is written down, with the result key pressed down only at the end of the calculation.

In operations of the $a(b+c)$ type, the part in brackets must be computed first. Multiplication follows only afterwards. If there is only one such term, there is no need for a memory but if there are several, the use of a memory or recording on paper is indispensable.

The sum (or difference) of products of the type $ab+cd+ef+-...$ can be calculated using the $\boxed{\mathrm{M+}}$ and or $\boxed{\mathrm{M-}}$ memories, or by recording on paper. There is a trick which makes such intermediate recording unnecessary: applying the following sequence of keys:

$$a_1 b_1+a_2 b_2+a_3 b_3+ ... =$$

$$= \boxed{a_1}\boxed{\times}\boxed{b_1}\boxed{\div}\boxed{a_2}\boxed{+}\boxed{b_2}\boxed{\times}\boxed{a_2}\boxed{\div}\boxed{a_3}\boxed{+}\boxed{b_3}\boxed{\times}\boxed{a_3} ...$$

because, for instance for three terms

$$a_1 b_1+a_2 b_2+a_3 b_3+\left[\frac{\left(\dfrac{a_1 b_1}{a_2}+b_2\right) a_2}{a_3}+b_3\right] a_3. \tag{5.60}$$

Iteration can be applied to determine functional values for which there is no key on the calculator.

For example, let us take the calculation of $\sqrt{a}$ on a calculator which has no key for $\boxed{\sqrt{\ }}$. Let us assume that the approximate value of $\sqrt{a}$ is a_1, with its correction (deviation from the real value is k). $\sqrt{a}$ in this case is sought in the following form:

$$\sqrt{a} = a_1+k. \tag{5.61a}$$

Squaring the equation, k being negligibly small, (cf. Section 5.1.3) gives

$$a = a_1^2+2a_1 k+k^2 \tag{5.61b}$$

which gives

$$k \cong \frac{a}{2a_1}-\frac{a_1}{2} \tag{5.61c}$$

and, substituted into (5.61a)

$$\sqrt{a} \cong \frac{a_1}{2} + \frac{a}{2a_1}. \tag{5.62}$$

This equation is the basis of iteration. If e.g. the value of $\sqrt{2}$ is to be calculated then, with a calculator which has an $\boxed{M+}$ calculating memory, the following iterative scheme may be used (for simplicity shown without the boxes around the symbols on the keys).

$$a_1 \div 2 (M+) a \div 2 = \div a_1 + (MR) =$$

(the result is fed back as the new a_1)

In the case of 2, with $a=2$, let us set out with $a_1=1$.

(a) 1 → 1.5

(b) 1.5 → 1.416 666 6

(c) 1.416 666 6 → 1.414 215 6

(d) 1.414 215 6 → 1.414 213 5

(e) 1.414 213 5 → 1.414 213 6 (unchanged).

Similar iteration would yield, e.g. $\sqrt[n]{a}$ where the iterative formula would be

$$a_2 \cong \frac{1}{n}\left[\frac{a}{a_1^{n-1}} + (n-1)a_1\right], \tag{5.63}$$

where a_1 is the first iterative value and a_2 the second (closer) iterative value.

Note: For iteration, calculators with memory should be used, to cope with calculations repeated with each new improved value.

Inverse Polish notation (IPN) has already been outlined in the section on the logic structure of calculators. Its more detailed treatment we postponed until now, after the treatment of the concepts of memory, register, etc. For the application of IPN a so-called stack memory [5.5] is required. These memories have several compartments, and during the operation which follows the entry of data, the information preceding the operation is transferred to the adjacent compartment. In Table 5.5 the reader may follow the calculation of the term $a \times b + c \times d$ with a four-compartment stack memory.

TABLE 5.5.

Contents of a four-compartment stack memory while executing the operation ab+cd using IPN

Number	Operation	Contents of stack			
		Display register	Non-display registers		
1	a (Data entry)	*a*	*a*		
2	b (Data entry)	*b*			
3	x (Operation)	*ab*			
4	c (Data entry)	*c*	*ab*		
5	d (Data entry)	*d*	*c*	*ab*	
6	x (Operation)	*cd*	*ab*		
7	+ (Operation)	*ab+cd*			

Programmable calculators

On the depression of keys programmable calculators (pocket size or desk-top) can store programs in their memory, and re-run them. The concept of programmability is very broad and since programmable calculators of different makes show greater dissimilarities than their non-programmable counterparts do, they cannot be described in general terms. In what follows, on the basis of [5.5], [5.7], [5.8], [5.9] and [5.10], we shall give a brief survey of the capabilities of programmable calculators, drawing the reader's attention to the relevant manuals and the instructions for use which accompany the different types.

According to [5.5], the most important instructions are as follows:

for programme editing:

STORE N	Store the data at the point *N*
RECALL N	Recall the data from the point *N*
LABEL M or MARK M	Label the step. The possible values of *M* are given on certain keys
GOTO M	Jump to label *M* unconditionally
IF (rel)	Conditional jump, depending on a result
SET FLAG	Adjust the marking character
CLEAR FLAG	Clear the marking character
CALL SR	Call a subroutine
RETURN	Jump back to the main program

To select the mode of operation:

DISP	Control of the display
RUN	Denotes several states, such as manual counting, running of program, handling peripherals

PRGM	Production and checking of program
WRITE	Writing a program
SWITCH	Program branch-off, depending on the position of the switch

With such a set of instructions a good number of different programs can be produced, although compared with higher-level programming languages it seems rather primitive. However, the great advantage, is that the calculator remains a handy instrument, a permanently available aid for technical people.

To test programs, the calculator must have a facility to move its program one step forward – preferably, one step in either direction. Most manufacturers provide this facility.

In some calculators the keys can be used as labels. This in turn enables the provision of "keys to be defined by the operator", which render frequently-used programs or their parts directly accessible.

Manufacturers generally include, with their programmable calculators, special calculator program packages for different purposes and different fields (e.g. the design of machines, electrical and electronic circuits, structural engineering projects, medical diagnostics, mathematics, probability and statistical calculations etc.). Out of these, the programs developed for the solution of special statistical problems (e.g. averaging; calculation of scatter, moments, and different distributions — normal, binomial, χ^2, Weibull, Student, etc. regression analysis; interpolation processes; determination of correlation coefficient; etc.) are of greatest significance for measurement evaluation.

Some calculators are larger and more sophisticated than the ones dealt with so far. They provide more possibilities in programming and allow the execution of very complicated and complex operations. The reason for classifying them with calculators is that they retain the most important characteristics of calculators, namely, control by keys.

Although the development of calculators has been unbroken for a long time now, it is far from finished.

Development is likely to continue in the following main directions

— the reduction of their physical size, or higher complexity with the same dimensions,
— reduction in price,
— a wider scope of problems which they can solve.

PART III

MEASUREMENT EVALUATION; MEANS AND METHODS

As shown in Fig. 2.1, the object measured yields so-called crude measurement information.

In the simplest case the measurement information is a single datum (measured datum — measured value) e.g. dimension, mass, temperature, etc. In other cases, measurement is performed in order to determine several data (e.g. temperature at selected points in a defined area, the determination of flow characteristics, etc.). Here the result is a group or a set of data. The individual data of measured sets may be mutually independent, or in functional relationship. In another even more complex case the measurement information is carried by a continuously (or discretely) varying signal, the measured signal, and not by a single or a finite number of measured data (measured values), e.g. in the measurement of the time function of a pressure varying with time, in recordings, digital records, etc.

The measurement information obtained at the information-indicating portion of the measuring device (cf. Fig. 2.1) (readouts, plots, recordings, etc.) is crude, because it has not been compared with the conditions under which the measurement took place. It is the task of measurement evaluation to produce physically correctly evaluated (sometimes corrected) data with properly estimated errors which can be regarded as the qualitative and quantitative properties of the measured object, on the basis of the crude measurement information. The aim of measurement and of measurement evaluation is never to obtain absolutely accurate information, but is rather to estimate the extent to which the result may, or does, deviate from the true value, i.e. to estimate the error in the measuring process.

In this third part of the book we will systematically survey the means and methods of evaluating crude measurement information. Chapters 6, 7, 8 will deal with the evaluation of deterministic and Chapter 9 with that of stochastic (random) measurement information.

6. EVALUATION OF DETERMINISTIC MEASUREMENT INFORMATION PART 1.

Evaluation of zero- and one-dimensional measurement information

6.1 GENERAL PROBLEMS

In Chapter 3 we discussed in considerable detail the dimension of measurement information, and stated that information is zero-dimensional if it contains no quantitative data and says only whether a measurand exists. We call measurement information one-dimensional when it contains a single quantitative datum i.e. the informative parameter.

Lacking numerical data, zero-dimensional measurement information cannot be numerically evaluated while for one-dimensional measurement information there are several distinct problems in measurement and evaluation.

Measurements fall into two major groups, notably measurements executed on single objects, and series of measurements of a single quantity on numerous (similar) objects.

One-dimensional measurement information may be acquired as:

— individual measurements,
— serial measurements of a single quantity,
— repeated measurements to enhance the accuracy of the single quantity obtained by measurement series,
— individual qualification instead of individual measurement.

The above cases may occur in the measurement of a single object or in the measurement of a single quantity on numerous objects.

Although still widely applied in industrial practice, individual measurement is far from ideal measuring technique. The only result a single measurement can provide is the readout of the value. The well known rule of thumb: "one measurement is no measurement" was the result of people realising this. A large number of individual measurements may be considered equivalent to measuring a single quantity on many (almost identical) objects. This might be justified if the objects are really identical (e.g. one dimension or one property of mass-produced components), and when the measuring process eliminates subjective errors (e.g. fully automated measurement).

A series of measurements of a single physical quantity datum means that in order to apply probability calculus in the estimation of the error, every measurement is repeated n times. This case is important because the result it yields can be applied to other cases as well.

The replacement of individual measurements by individual qualification means measurements with a gauge, and consists of reducing the acquisition of one-dimensional measurement information to the acquisition of zero-dimensional measurement information. This reduction is only virtual because by the designation of two limits (LL lower limit and UL upper limit) the workpieces can be divided into two groups in such a way that one group includes those which fall between the limits and those which fall outside the limits.

6.2 EVALUATION OF ONE-DIMENSIONAL MEASUREMENT INFORMATION

One-dimensional measurement information contains a single numerical (quantitative) datum, the informative parameter: length, mass, duration, current, temperature, etc. Since a single measurement does not yield information to which statistical or probability laws can be applied, the measurement is repeated. Accordingly, if measurement of the same quantity or numerous (similar) objects (e.g. mass-produced parts) is involved, by virtue of the character of manufacture and in-plant supervision, probability and statistical methods can be resorted to [6.1], [6.2], [6.3].

6.2.1 MEASUREMENT SERIES, SAMPLES OF n ELEMENTS. THEIR PROBABILITY AND THEIR STATISTICAL CHARACTERISTICS

Measurement series

Let us assume that a dimension or some other measured data has been measured n times under identical conditions, so as to allow the application of probability calculus in the estimation of the random errors. The set of measured values so obtained is called a measurement series with n elements, or simply a measurement series. Measurement series can be evaluated by the methods of probability calculation.

Sample of n elements

Let us assume that the task is to establish a dimension or other property of elements in a large set of objects (e.g. some of the dimensions of a workpiece produced in very large runs). Since in practice it is impossible to execute the measurement on every piece of the set, the measurement is executed on a number n, arbitrarily picked from the set, and from the result conclusions are drawn as to the measured data concerned.

The set of objects being measured together with the measurands in question are called a statistical set (briefly, set), and the n elements arbitrarily selected are called a sample of n elements (briefly, sample). On the basis of the measured properties of the sample, we may draw conclusions as to the properties of the elements in the set (even the elements not measured) by means of mathematical statistics.

Summing up, it may be stated that although not identical, the measurement series of n elements and the sample of n elements are allied concepts. Due to their many common properties we shall deal with them together – as far as possible – and draw the reader's attention to any dissimilarities.

Measurement series and their probability characteristics

Let x be the measured value of the measurand. The ith member of the measurement series will be x_i, its true (correct) value x_0 and its undefined but probable (expected) value $\bar{x}$.

The absolute error is the difference between the measured value and a reference value, the relative error the value of the absolute error compared with the true or expected value of the measurand. The units of the absolute error are the same as those of the measured physical quantity, while the relative error is a pure number. Both the absolute and the relative errors may be apparent or true, depending on whether the expected value $\bar{x}$ or the theoretically more important true value x_0 has been used as a reference in arriving at the difference. Denoting the absolute errors by Δ and the relative errors by δ the virtual absolute error is

$$\Delta x_i = x_i - \bar{x} \tag{6.1}$$

and the actual absolute error

$$\Delta x_{0i} = x_i - x_0, \tag{6.2}$$

the apparent relative error

$$\delta x_i = \frac{x_i - \bar{x}}{\bar{x}} = \frac{\Delta x_i}{\bar{x}}, \tag{6.3}$$

the true relative error

$$\delta x_0 = \frac{x_i - x_0}{x_0} = \frac{\delta x_{0i}}{x_0}. \tag{6.4}$$

The expected value (the most probable value) follows from the criterion which demands that the square of the errors should be minimal,

$$\sum_{i=1}^{n} (x_i - \bar{x})^2 = \sum_{i=1}^{n} x_i^2 = \text{minimum!} \tag{6.5}$$

The minimum can be obtained by the calculation of the extreme value, and the expected value is the arithmetic mean of the measured values:

$$\bar{x} = \frac{x_1 + x_2 + \ldots + x_n}{n} = \frac{1}{n} \sum_{i=1}^{n} x_i . \tag{6.6}$$

For similar measurements executed under identical conditions, the normal (Gaussian) distribution usually applies. We accept this assumption but, in accordance with our earlier notation, instead of the discrete variables x_i, we switch to a continuous distribution, which yields a density function:

$$f(x) = \frac{1}{\sigma\sqrt{2\pi}} \exp\left\{-\frac{(x-\bar{x})^2}{2\sigma^2}\right\}, \tag{6.7}$$

where $\bar{x}$ the expected value, the so-called centre, of the distribution, is real and where σ the standard deviation must be positive.

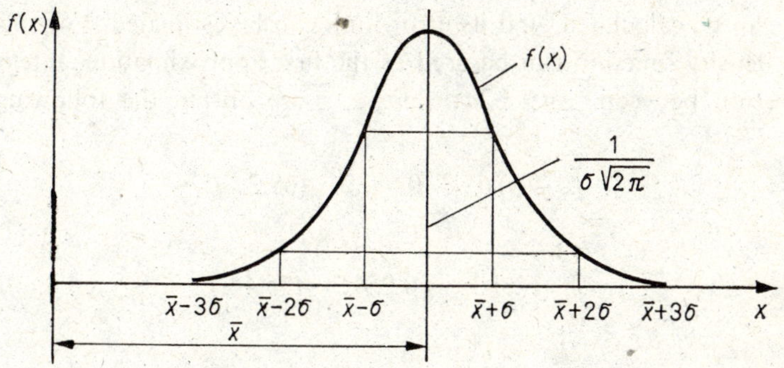

Fig. 6.1. *Density function for a measurement with a Gaussian curve*

Figure 6.1 shows the density function, the Gaussian curve. Standard deviation, as we know, is the positive square root of the mean of the square of the deviation from the expected value, calculated for $(n-1)$ members of the series

$$\sigma = +\sqrt{\frac{1}{n-1}\sum_{i=1}^{n}(x_i-\bar{x})^2}. \tag{6.8}$$

Remember that $(n-1)$ is called the degree of freedom of the distribution. If $n \gg 1$ then $(n-1) \cong n$, and

$$\sigma \cong +\sqrt{\frac{1}{n}\sum_{i=1}^{n}(x_i-\bar{x})^2}. \tag{6.9}$$

From the density function of the probability distribution, i.e. on the data it yields, various types of error can be derived. The basic definition of the error was given in (6.1) to (6.4) inclusive, and some others will be given below.

The simple mean error or average error, also called the expected deviation, is the average of the absolute values of the individual errors. Using the bar over the symbol to denote the average, we have

$$\overline{|\delta x|} = \frac{1}{n}\sum_{i=1}^{n}\left|\frac{x_i-\bar{x}}{\bar{x}}\right| = \frac{1}{n}\sum_{i=1}^{n}|\delta x_i|. \tag{6.10}$$

On the basis of the density function (6.7) we may calculate that

$$\Sigma|\delta x_i| = 0.798\sigma. \tag{6.11}$$

The mean square error is the standard deviation proper, as in (6.8), to which we shall revert later on.

The probable error (ε) is a quantity which makes it probable that the absolute value of the error be smaller than ε, $\frac{1}{2}$. From the density function (6.7) it follows that

$$\varepsilon = 0.675\sigma. \tag{6.12}$$

Using the above relationship the expected value of a measuring series consisting of n members can be calculated, and its error limit can be estimated. For such an estimation the density function can be used as the first approximation. Integrating the density function between $\pm\sigma$; $\pm 2\sigma$; $\pm 3\sigma$; $\pm\infty$ we obtain the following results:

$$\int_{-\sigma}^{+\sigma} f(x)\,\mathrm{d}x = 0.6823 \quad (68.2\%); \tag{6.13a}$$

$$\int_{-2\sigma}^{+2\sigma} f(x)\,\mathrm{d}x = 0.9542 \quad (95.4\%); \tag{6.13b}$$

$$\int_{-3\sigma}^{+3\sigma} f(x)\,\mathrm{d}x = 0.9973 \quad (99.7\%); \tag{6.13c}$$

$$\int_{-\infty}^{+\infty} f(x)\,\mathrm{d}x = 1.000 \quad (100\%), \tag{6.13d}$$

to be evaluated in the following way:

— if we assume that the uncertainty of the measurements is $\pm\sigma$ then we have to accept that only 68% of the measurements would show an error of less than $\pm\,\sigma$ and the balance of 32% would be even less accurate;
— if the uncertainty is taken to be $\pm 2\sigma$ then 5% of the measurements would still exceed the chosen error limit of $\pm 2\sigma$;
— if the uncertainty is taken to be $\pm 3\sigma$ then 99.7% of the measurements, i.e. practically every one, would fall into this zone and these boundaries may be regarded as the error limits.

The value of σ depends on the measuring instrument, the measurement conditions, the person executing the measurements, the number of measurements executed, etc. With other σ values the density function (error curve) will assume different shapes (Fig. 6.2).

If the standard deviation is considerable, the curve will be flat and the $\pm 3\sigma$ error limit wider, while if σ is smaller, the curve will be narrower and steeper.

This error curve is plotted on the basis of n measurements i.e. a measurement series of n members. The maximum deviation of a single measurement from the expected value may be $\pm 3\sigma$.

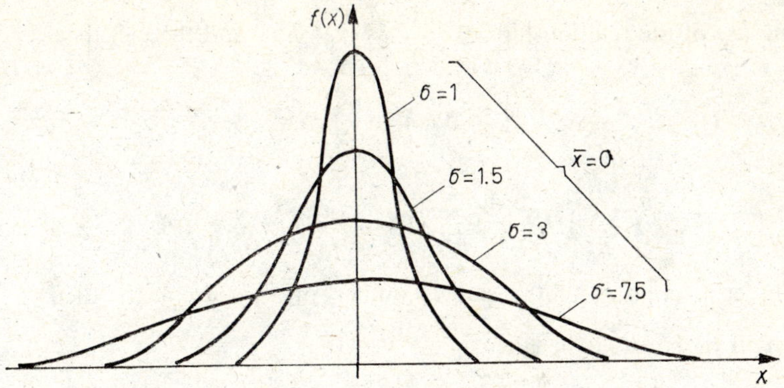

Fig. 6.2. *The influence of different values of the standard deviation on the density function (error curve)*

Increasing the accurary of a measurement series by increasing the number of measurements

On the basis of the above, it may be assumed that with an increasing number of measurements the results may become more accurate. Equation (6.6) established the arithmetic mean as the most likely value of the measurand. Its uncertainty was given by σ.

If the arithmetic mean is determined from several series of measurements, its values will probably be different. The arithmetic mean of these new values may again be determined. Using the bar over the symbol as before and assuming a series of R measurements we have

$$\bar{\bar{x}} = \frac{\bar{x}_1+\bar{x}_2+\ldots+\bar{x}_k}{k} = \frac{1}{k}\sum_{j=1}^{k}\bar{x}_j .$$

Calculating the square of the mean error h of the series of arithmetic means:

$$h^2 = \frac{(\bar{x}_1-x_0)^2+(\bar{x}_2-x_0)^2+\ldots+(\bar{x}_k-x_0)^2}{k} = \frac{1}{k}\sum_{j=1}^{k}(\bar{x}_j-x_0)^2.$$

The individual members $\bar{x}_j$ – themselves being arithmetic means – are obtained by calculating the arithmetic mean of the individually measured values x_i of some jth series:

$$\bar{x}_j = \frac{1}{n_j}\sum_{i=1}^{n_j} x_i,$$

$$\bar{x}_j - x_0 = \frac{1}{n_j}\sum_{i=1}^{n_j}(x_i-x_0),$$

Substituted into the equation which defines h^2:

$$h^2 = \frac{1}{k}\sum_{j=1}^{k}\frac{1}{n_j^2}\Big[\sum_{i=1}^{n_j}(x_i-x_0)\Big]^2.$$

Making use of the relationship $\sigma^2 \cong \frac{1}{n} \sum_{i=1}^{n} (x_i - x_0)^2$ and the fact that

$$(\sum_{i=1}^{n} \Delta x_{0i})^2 \cong \sum_{i=1}^{n} \Delta x_{0i}^2$$

we have

$$h^2 = \frac{1}{k} \sum_{j=1}^{k} \frac{1}{n_j^2} \sum_{i=1}^{n_j} (x_i - x_0)^2 = \frac{1}{k} \sum_{j=1}^{k} \frac{1}{n_j} \sigma_k^2 .$$

If each series consists of the same number of measurements then $\frac{1}{n_j} = \frac{1}{n} =$ $=$const. will be constant in each series and so

$$h^2 = \frac{1}{n} \frac{1}{k} \sum_{j=1}^{k} \sigma_k^2 = \frac{1}{n} \bar{\sigma}^2 .$$

Thus, we may write, by approximation, that

$$h^2 = \frac{1}{n} \sigma^2 .$$

Taking square roots of both sides of the equation and denoting the mean uncertainty in the arithmetic mean by u we obtain

$$u = \sqrt{h^2} = \pm \frac{\sigma}{\sqrt{n}} . \tag{6.14a}$$

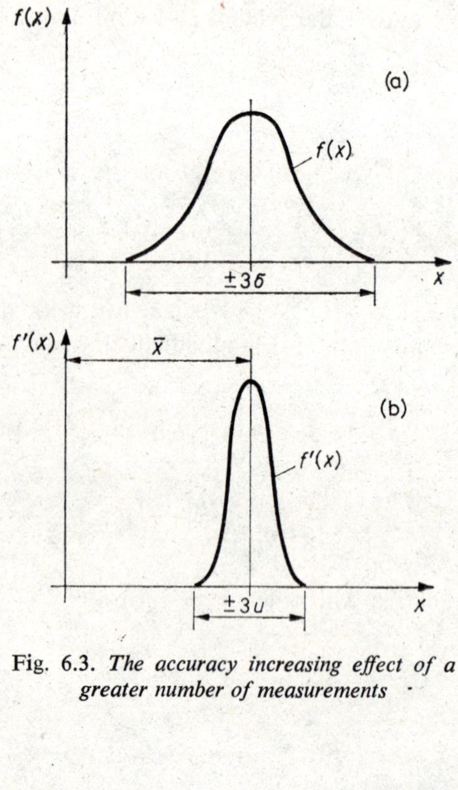

Fig. 6.3. *The accuracy increasing effect of a greater number of measurements*

This result may also be evaluated as implying that on increasing the number of measurements a new distribution with standard deviation of u will be obtained. To refine the distribution having a standard deviation of σ determined from n measurements, on executing the measurement n' times the new smaller standard deviation will be

$$u = \frac{\sigma}{\sqrt{n'}}. \tag{6.14b}$$

This means that in the measurement series the uncertainty is $\pm 3u = \pm 3\sigma/\sqrt{n'}$ and it includes also the case of $n' = 1$. The effect of the repetition of the measurements is shown in Fig. 6.3.

If the distribution of the standard deviation and the degree to which the accuracy must be increased are known, the rearranged form of (6.14) yields the required number of repetitions:

$$n' = \left(\frac{\sigma}{u}\right)^2. \tag{6.14c}$$

6.2.2 PRACTICAL EVALUATION

In Section 6.2.1 we briefly surveyed some basic problems in the evaluation of measurement series on the basis of probability. These principles and basic notions are easily applicable in practice, provided that the number of elements, n, in the measurement, either series or samples, is relatively small, not greater than 15 to 20. However, the situation in practice is different. The number of measurements is usually considerable: it may run into hundreds or even thousands. Thus, even the production of a simple mean value according to (6.6) may call for a few hundred measurements. The handling of large sets of data is difficult and causes errors in the calculation.

This chapter is devoted to methods of handling large quantities of data which may occur in practice, ways for their possible reduction, and the numerical or graphical operations that may be involved. In the treatment of this subject we shall follow Section 6.2.1, but also discuss relationships which facilitate practical work (approximations, reductions etc.).

We set out once more from a sample, i.e. a measurement series of n elements where the lowest and highest values are $\check{x}$ and $\hat{x}$.

To reduce the number of data, it is advisable to classify the data x_i (the elements of the sample i.e. measurement series) into a number k of interrelated intervals. Classification here means arrangement by size. Let the number of classes be k, whose value can be chosen according to the empirical rule $k \cong \sqrt{n}$* but should preferably be

* For the determination of the number of classes the literature recommends another empirical rule (formulated by Struges) which says that $k \cong 3.3 \log n$ to base 10.

between 5 and 25. Let us attach a number $m=1, 2, \ldots, k$ to each class, and call it the class number. Assuming that each class contains p elements (measured data), we may write that $kp \leqq n$. (We draw the reader's attention to the fact that p may assume different values.) If a certain element, in our case a certain characteristic measured value x_m, occurs h times in a measuring series (sample), then $h(x_m)$ is the absolute frequency of the value x_m:

$$h(x_m) = h_m = h. \tag{6.15}$$

It is important that

$$\sum_{m=1}^{k} h_m = n, \tag{6.16}$$

where k denotes the possible number of measured values and n the number of all measured values.

If the value x_m occurs h times in a measuring series (sample) then the ratio of the absolute frequency of $h(x_m)$ to the number of elements in the series (sample) will be the relative frequency:

$$f(x_m) = \frac{h(x_m)}{n} = f_m \tag{6.17}$$

and the values of f_m – like the values of relative frequency in general – lie between zero and 1.

The percentage value of the relative frequency is the % (percentage) frequency

$$f(x_m)\% = \frac{h(x_m)}{n} 100\%. \tag{6.18}$$

The sum of the percentage frequencies is the total percentage frequency

$$F_m = \sum_{m=1}^{k} f(x_m)\% = f_1 + f_2 + \ldots + f_k. \tag{6.19}$$

With these data, it is possible to determine the frequency distribution. A series of numbers, a table or a graph indicates the frequency in the different classes of the measuring series (sample). In practice the frequency distribution of the measuring series (sample) is illustrated by

— tally chart,
— line chart (histogram),
— linked-line chart.

In addition to the above, the stepped chart is suitable for the illustration of the distribution of the percentage total frequency.

Here is an example to show the methods of plotting and using such diagrams.

Example

Take a group of mass-produced workpieces, consisting of $n=106$ pieces. Measuring a certain dimension on each, depending on the purpose of measurement, we shall have a measuring series (sample) consisting of $n=106$ members, i.e. a sample of $n=106$ elements. The measured values show a statistical distribution between 16.049 and 16.060 mm. The difference between the lowest and the highest values of the series (sample), i.e. between $\check{x}=16.049$ and $\hat{x}=16.060$ is called the range (R) of the series (sample)

$$R = \hat{x} - \check{x} = 16.060 - 16.049 = 0.011 \text{ mm.}$$

The range is equal to 11 microns. In classification, it is advisable to divide the range into equal parts in such a way that the mid-point of each division is a measured value. The case will be easy if the range is divided into microns, since the measurement was made with an accuracy of 1 micron.

In this way, each measured value will be equal to the chosen mid-point of a class. (This is not necessarily so; consider what the result would be if the same measurement series were made with an accuracy of 0.1 micron – i.e. four decimal places of mm.)

In each interval (class) in the divided range the tally chart contains as many marks as there are measurement results which fall into the interval concerned. Tally charts are made using a so-called original register which contains every piece of measurement information in the order over time in which they collected. Figure 6.4 shows a tally chart. If the tally chart illustrates also the tolerance limits determined before the measurement, it will allow conclusions to be drawn. For instance in Fig. 6.4, the following conclusions can be drawn: the tolerances were adhered to but the average value has shifted towards the smaller end.

On the basis of the tally chart line charts and frequency charts can be plotted. They are essentially equivalent to line lists but more illustrative. The tally chart contains the number of lines only, the charts also contain the measured true-to-scale ordinates

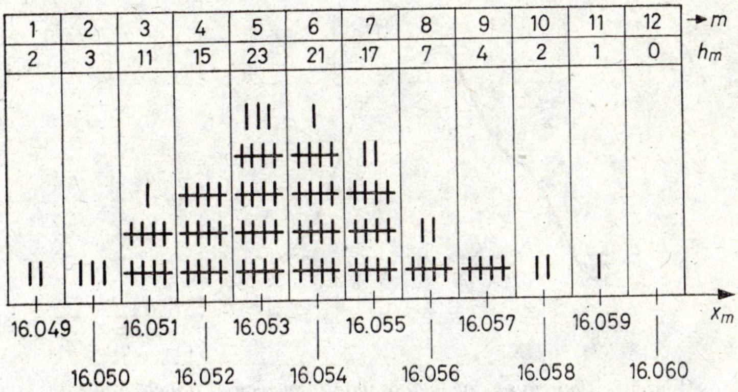

Fig. 6.4. *Tally chart*

of the same numbers. Figure 6.5 illustrates the three charts mentioned in the example (all three may be regarded as empirical variants of the probability density function shown in Fig. 6.1) and shows also that the three charts are essentially equivalent.

Figure 6.6 shows the so-called stepped chart. Its abscissa corresponds to those of the three graphs mentioned before, but on its ordinate the total percentage frequency according to (6.18) is also indicated.

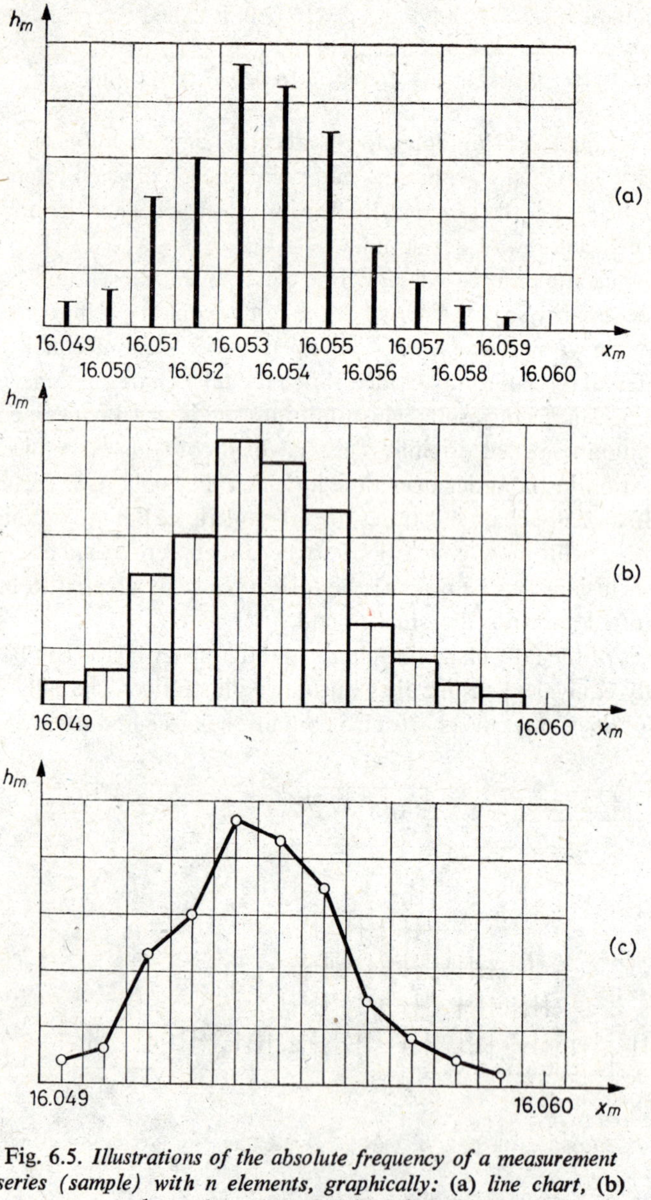

Fig. 6.5. *Illustrations of the absolute frequency of a measurement series (sample) with n elements, graphically:* (a) *line chart,* (b) *column chart (histogram)* (c) *linked line*

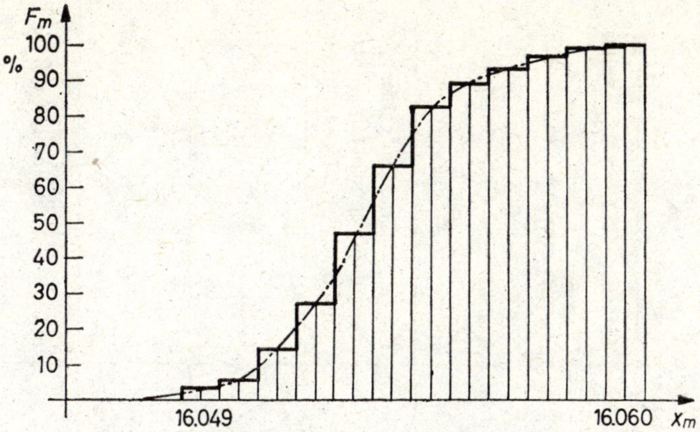

Fig. 6.6. *Stepped chart of percentage total frequency (the abscissa is the same as that in Fig. 6.5)*

Now let us return to the practical evaluation of the measurement series (sample) consisting essentially of the determination of the mean value and the standard deviation. On their basis, it will be possible to estimate the "safety" of the results arrived at by statistical means.

We shall deal first with the different practical ways of determining the mean (expected) value according to (6.6), because calculation from the definition with a large number of data would be difficult and time-consuming. First a number of auxiliary quantities are defined. The range R has been dealt with in the example, and its mid-point is

$$R_k = \frac{\hat{x}+\check{x}}{2}, \tag{6.20}$$

which may differ quite considerably from the arithmetic mean.

On the other hand the difference between the median and the arithmetic mean according to (6.6) is usually negligibly small. The median ($\tilde{x}$) is a kind of central value which, with an odd number of n, is equal to the middle measured value while with an even number of n, it is equal to the arithmetic mean between the two middle values.

The mode (modal value or mean density) m_d is the most frequently occurring value in the set of classified values (in our example $m=5$, and $m_d=16.053$ mm).

The geometric mean is

$$x_g = \sqrt[n]{\prod_{i=1}^{n} x_i} \tag{6.21}$$

and, like the arithmetic mean (6.6), it depends on each value of x_i. The effect of the extreme values $\check{x}$ and $\hat{x}$ is less than that of those values on the mode m_d. If the measured values are equal, then $x_g \leqq \bar{x}$ and $x_g = \bar{x}$.

The method of calculating the arithmetic mean (the expected value) from the measured data, according to Hofmann [15], have been compiled in Table 6.1. From the table the method best suited for the case in hand, can be selected.

TABLE 6.1.

Methods of calculating the arithmetic mean [15]

Name of the method	Method of calculating the arithmetic mean	Amount of work involved in the calculation	Comment
By definition	$\bar{x} = \frac{1}{n} \sum_{i=1}^{n} x_i$ x_i is a general member of the series (sample)	Large	The size of the series (sample) if $n \leqq 25 \cdot n > 25$ pocket calculator is advisable
Frequency method	$\bar{x} = \frac{1}{n} \sum_{m=1}^{k} h_m u_m$ h_m is the absolute frequency, u_m is the middle value of the class	Small	If $n > 25$ then a tally chart is advisable first
Calculation with the auxiliary quantity x_s	$\bar{x} = \bar{x}_s + \frac{1}{n} \sum_{i=1}^{n} s_i$ $s_i = x_i - \bar{x}_s$, x_s is the assumed mean value	Large	For large numerical values; for quality control in production only under certain conditions
Tally chart with the use of graph papers	$\bar{x} = \bar{x}_s + \frac{d}{n} \sum_{m=1}^{k} m h_m$ m is the class number, $\bar{x}_s$ is the assumed mean value, d is the width of class	Small	Using mm graph paper
Estimate of the expected value	1. $\bar{\bar{x}} = \frac{1}{k} \sum_{i=1}^{k} \bar{x}_i = \mu$ $\bar{x}_i$ the arithmetic mean of the ith class	Large	Number of classes $k \leqq 25$
	2. $\bar{\tilde{x}} = \frac{1}{k} \sum_{i=1}^{k} \tilde{x}_i = \mu$ $\tilde{x}_i$, the median of the ith class	Small	If $k = 3$ or 5, estimation will be simple
	3. $\bar{R}_k = \frac{1}{k} \sum_{i=1}^{k} R_{ki}$ R_{ki} the midpoint of the range of the ith class	Small	Advantageous in practice

Summing up, to evaluate a measurement series with n elements its expected value and its scatter must be determined. In a determination on the basis of the original register, usually some other quantities (range, median, mode, etc.) may be involved and it is generally necessary to plot the tally chart (Fig. 6.4) first, and then on its basis some of the charts in Fig. 6.5. This information, collated with the conditions under which the measurement was carried out (the instrument, the environment, the operator, the conditions of manufacture) is necessary for the complex process of evaluation. For the execution of the process no general method can be recommended due to the great variation of possible conditions. It will always depend on the case in hand.

All this, however, will not complete the task we have with samples of n-elements. To arrive at further conclusions from the data obtained, the method of mathematical statistics is required. From the expected value of the measurement series, its standard deviation, etc., it is possible to judge the accuracy and reliability of the measurement results with a fair degree of certainty but their values are not identical to the expected value and the standard deviation of the set.

Let us therefore assume a statistical set of N elements and a sample of n elements derived from it. The data in N are called theoretical data, while the data in the sample of n elements are known as empirical data (theoretical distribution, empirical distribution). Mathematical statistics approximates the theoretical data to the empirical ones.

The most important empirical data of the sample of n elements $x_1, x_2, \ldots, x_i, \ldots, x_n$ yields the empirical expected value:

$$\bar{x} = \frac{x_1+x_2+\ldots+x_n}{n} = \frac{1}{n}\sum_{i=1}^{n} x_i. \tag{6.22}$$

The empirical variance

$$s^2 = \frac{(x_1-\bar{x})^2+(x_2-\bar{x})^2+\ldots+(x_n-x)^2}{n} = \frac{1}{n}\sum_{i=1}^{n}(x_i-\bar{x})^2. \tag{6.23}$$

Let us assume that the theoretical distribution of the set has an expected value and a standard deviation denoted by m and σ respectively. Since

$$M(\bar{x}) = \frac{1}{n} M(x_1+x_2+\ldots+x_n) = \frac{1}{n} nm = m, \tag{6.24}$$

moreover, according to the law of averages, with higher n, $\bar{x} \to m$. Let the distribution function of the set be $F(x)$ and its probability variable ξ. By definition, the standard deviation of the set is

$$\sigma = D(\xi). \tag{6.25}$$

The expected value of the empirical variance is not equal to the theoretical variance. We will not go into the verification, but

$$M(s^2) = \frac{n-1}{n}\sigma^2, \tag{6.26}$$

therefore, to approximate σ^2 instead of s^2 the so-called corrected empirical variance

$$s^{*2} = \frac{n}{n-1}\sigma^2 = \frac{1}{n-1}\sum_{i=1}^{n}(x_i - \bar{x})^2 \tag{6.27}$$

can be used where

$$M(s^{*2}) = \sigma^2. \tag{6.28}$$

Now let us see the correlations between the data relating to the sample and those relating to the set. Let us assume that the set has a normal distribution, an expected value m and standard deviation σ. In such cases we say that the set has a distribution of $\mathfrak{N}(m, \sigma)$. This assumption usually holds good for practical purposes. Since $\bar{x}$ is the sum of independent, normally distributed probability variables, it is itself normal and its distribution $\mathfrak{N}(n, \sigma/\sqrt{n})$.

It follows – without verification here – that with a set of $\mathfrak{N}(m, \sigma)$ distribution both $\frac{n}{\sigma^2}s^2$ and $\frac{n-1}{\sigma^2}s^{*2}$ are probability variables of a distribution of χ^2. (As regards the distribution χ^2 we refer to Chapter 4 and to textbooks on probability calculus and mathematical statistics, e.g. [4.1] and [4.2]).

In the case quoted it was assumed that σ was known, although in practice we know or assume only that the set is normally distributed, not the value of σ. In such cases it

TABLE 6.2.

The t_p values of the Student distribution for the determination of the confidence interval of the expected value

Degrees of freedom \ Value of p	Values of t_p				
	0.001	0.01	0.05	0.1	0.5
1	636.619	63.657	12.706	6.314	1.000
2	31.941	9.925	4.303	2.920	0.816
3	12.941	5.841	3.182	2.353	0.765
4	8.610	4.604	2.776	2.132	0.741
5	6.589	4.032	2.571	2.015	0.727
6	5.959	3.707	2.447	1.943	0.718
7	5.405	3.499	2.365	1.895	0.711
8	5.041	3.355	2.306	1.860	0.706
9	4.781	3.250	2.262	1.833	0.703
10	4.587	3.169	2.228	1.812	0.700
15	4.073	2.942	2.131	1.753	0.691
20	3.850	2.845	2.086	1.725	0.687
25	3.725	2.787	2.060	1.708	0.684
30	3.646	2.750	2.042	1.697	0.683
40	3.551	2.704	2.021	1.684	0.681
60	3.460	2.660	2.000	1.671	0.679
120	3.373	2.617	1.980	1.658	0.677
∞	3.291	2.576	1.960	1.645	0.674

is important to estimate the so-called limits of validity – tolerance limits – of m and σ i.e. the range within which the approximation of m by $\bar{x}$ and of σ by s^* the deviations $\bar{x}-m$ and $s^*-\sigma$ will fall into a given interval with a known degree of certainty. Mathematical statistics calls this random interval the "confidence interval", and its upper and lower limits are known as confidence limits.

Again, without verification, we state that in the case of a set of the distribution of $\mathfrak{N}(m, \sigma)$ the confidence interval for m is as follows

$$P\left(\bar{x}-t_p\frac{s^*}{\sqrt{n}} \leqq m \leqq \bar{x}+t_p\frac{s^*}{\sqrt{n}}\right) = 1-p \tag{6.29}$$

and for σ

$$P\left(\frac{ns^2}{\chi^2_{p/2}} \leqq \sigma^2 \leqq \frac{ns^2}{\chi^2_{1-p/2}}\right) = 1-p. \tag{6.30}$$

According to the above relationships the probability that m will not differ from $\bar{x}$ by more than $\pm t_p\frac{s^*}{\sqrt{n}}$, and that σ falls between the confidence limits of $\frac{ns^2}{\chi^2_{p/2}}$ and $\frac{ns^2}{\chi^2_{1-p/2}}$ is $1-p$ with $(1-p)\times 100\%$ being the known confidence level. In practice the

TABLE 6.3.

The values of the χ^2_p distribution for the calculation of the confidence limits of the variance (standard deviation)

Degrees of freedom \ Values of p	Values of χ^2_p							
	0.001	0.01	0.02	0.05	0.10	0.30	0.95	0.99
1	10.827	9.210	5.412	3.841	2.706	1.074	0.003	0.000
2	13.815	9.210	7.824	5.991	4.605	2.408	0.103	0.020
3	16.268	11.345	9.837	7.815	6.251	3.665	0.352	0.115
4	18.465	13.277	11.668	9.488	7.779	4.878	0.711	0.297
5	20.517	15.086	13.388	11.070	9.236	6.064	1.145	0.554
6	22.457	16.812	15.033	12.592	10.645	7.231	1.635	0.872
7	24.322	18.475	16.622	14.067	12.017	8.383	2.167	1.239
8	26.125	20.090	18.168	15.507	13.362	9.524	2.733	1.646
9	27.877	21.666	19.679	16.919	14.684	10.656	3.325	2.088
10	29.588	23.209	21.161	18.307	15.987	11.781	3.940	2.558
12	32.909	26.217	24.054	21.026	18.549	14.011	5.226	3.571
15	37.697	30.578	28.259	24.996	22.307	17.322	7.261	5.229
18	42.312	34.805	32.346	28.869	25.989	20.601	9.390	7.015
20	45.315	37.566	35.020	31.410	28.412	22.775	10.851	8.260
22	48.268	40.289	37.659	33.924	30.813	24.939	12.338	9.542
25	52.620	44.314	41.566	37.652	34.382	28.172	14.611	11.524
28	56.793	48.278	45.419	41.337	37.916	31.391	16.928	13.565
30	59.703	50.892	47.962	43.773	40.256	33.530	18.493	14.953

most frequent confidence levels are 90, 95 and 99% corresponding to $p=0.1, p=0.05$ and $p=0.01$ respectively. t_p, $\chi^2_{p/2}$ and $\chi^2_{1-p/2}$ can be obtained from the Student and χ^2 distribution tables (Tables 6.2 and 6.3). In relation to Eq. (6.30), we note that the values of $\chi^2_{p/2}$ and $\chi^2_{1-p/2}$ may not be directly read from the table because e.g. for $p=0.05$, $p/2=0.025$ and $1-p/2=0.975$ do not occur in the table.
In such cases linear interpolation must be applied.

Example

Grooves of 5.12 H8 are cut into workpieces on a machine equipment. In the ISO system this corresponds to $5.12_0^{+0.018}$, with tolerances. From the large series a sample with 16 elements is picked out and the size of the grooves measured, one by one. The results were: 5.121; 5.123; 5.124; 5.131; 5.122; 5.123; 5.132; 5.131; 5.124; 5.126; 5.128; 5.127; 5.133; 5.129; 5.125; 5.129 mm. Averaging gave the following expected value:

$$\bar{x} = \frac{1}{n}\sum_{i=1}^{n} x_i = \frac{1}{16}(5.121+5.123+\ldots+5.129) = \frac{82.028}{16} = 5.126\,75 \text{ mm.}$$

From this the empirical variance can be calculated

$$s^2 = (1/n)[(5.121-5.126)^2+(5.123-5.126)^2+\ldots+(5.129-5.126)^2] = 1.356\times10^{-5}.$$

The empirical standard deviation

$$s = \sqrt{s^2} = \sqrt{1.356\times10^{-5}} = 3.68\times10^{-3},$$

the corrected empirical variance

$$s^{*2} = \frac{n}{n-1}s^2 = \frac{16}{15}\times1.356\times10^{-5} = 1.446\times10^{-5},$$

and the corrected standard deviation

$$s^* = \sqrt{1.446\times10^{-5}} = 3.802\times10^{-3}.$$

From the data obtained from the 16-element sample, let us now draw conclusions about the set, the pieces produced during the life of the tool edge, with a confidence level of approximately 90%, corresponding to $p=0.1$. To calculate the confidence interval referring to the expected value the value of t_p can be taken from Table 6.2, with degrees of freedom $n-1=16-1=15$. With $p=0.1$ the table shows that $t_p=1.753$ and the confidence limit is

$$t_p\frac{s^*}{\sqrt{n}} = 1.753\times\frac{3.802\times10^{-3}}{16} = 1.66\times10^{-3}\text{ mm} = 1.66\ \mu\text{m}$$

and the expected value $\bar{x}=(5.126\pm0.0016)$ mm.

To calculate the confidence limits of standard deviation, the data must be taken from Table 6.3 with $p=0.1$, $p/2=0.05$ and $1-p/2=0.95$. To these values $\chi^2_{0.05}=24.996$ and $\chi^2_{0.95}=7.261$ correspond.

$$\frac{ns^2}{\chi^2_{0.95}} = \frac{16\times 1.356\times 10^{-5}}{24.996} = 0.8679\times 10^{-5},$$

$$\frac{ns^2}{\chi^2_{0.05}} = \frac{16\times 1.356\times 10^{-5}}{7.261} = 2.988\times 10^{-5}.$$

Accordingly,

$$0.86\times 10^{-5} < \sigma^2(\cong s^{*2} = 1.446\times 10^{-5}) < 2.988\times 10^{-5}$$

which will tell us the magnitude of the uncertainties in the calculation.

In the example the empirical variance was calculated by (6.23). Calculations may be simplified and speeded up, particularly if a calculator is available, by using the formula

$$s = \sqrt{\frac{1}{n-1}\sum_{i=1}^{n} x_i^2 - \frac{\left(\sum_{i=1}^{n} x_i\right)^2}{n}}\,.$$

7. EVALUATION OF DETERMINISTIC MEASUREMENT INFORMATION. PART 2.

Evaluation of two- and multi-dimensional measurement information (signals) by value. Mathematical (numerical) and geometric interpretation of functional relationships determined by measurement

In the case of two- and multi-dimensional measurement information the measurand is a function of one or several physical quantities. This chapter will outline the different possibilities for their evaluation. For simplicity, we shall concentrate on the evaluation of two-dimensional measurement information, and regarding multi-dimensional information we shall refer to the two-dimensional case and the possibilities of reduction.

How to formulate the problem?

In the case of two-dimensional measurement information (function with a single variable) we assume that the result is obtained as $n+1$ interrelated pairs of numbers (x_i, y_i) in the interval $a=x_0<x<x_n=b$ of the independent variable x. If so, the function $y(x)$ determined by measurement is regarded as tabulated. Since, however, data compiled in tables contain errors and, since the discrete points $y_i(x_i)$ do not unambiguously determine $y(x)$, the approximate value of the function must be determined from the table with the required precision. Three- or multi-dimensional information further complicates the situation but by designating parameters, plotting sets of curves or, if necessary, by nomograms, we can illustrate them in the plane.

Methods of solution

Numerous methods can be applied to carrying out the basic task outlined above and all fall into one of the two basic categories

— algebraic methods which yield approximate, so-called empirical, formulae;
— geometric or graphical methods whose result is a graphical illustration, a curve, a nomogram, etc.

These will be discussed in the present chapter.

7.1 EMPIRICAL FORMULAE

In setting up empirical formulae the aim is to replace the functional relationships obtained by measurement by a function (algebraic or transcendent), whose parameters have been determined on the basis of the measured data, and in which the approximation is based on the estimated error which the data carry.

To set up empirical formulae, the following processes are used:

— interpolation,
— equalization,
— selection of the best approximate formula.

7.1.1 INTERPOLATION AND EQUALIZATION

The simplest and most frequently used methods for the analytical production of tabulated functions are interpolation and equalization.

Interpolation

If the values of the $n+1$ $y(x_i)=y_i$ functions are known, the unknown function $y=y(x)$ can be approximated by a suitable interpolation formula in the form $Y(x)\equiv Y(x, A_0, A_1, \ldots, A_m)$ in which the $n+1$ A_k parameters must be so determined that they satisfy the $n+1$ argument values: $Y(x_i)=y(x_i)=y_i$. The function of "known form" is mostly a polynomial, in some cases a trigonometric polynomial, less frequently a rational fractional function.

Equalization

Approximation by interpolation, particularly with polynomials is not always justified. With interpolation "exact" functional values are obtained about points which we know are not exact, i.e. contain errors. With the analytical construction of a tabulated function we may wish to compensate for the fluctuations of the $y(x_i)$ values caused by random errors. We therefore construct the same function $Y(x)\equiv Y(x, A_0, A_i, \ldots$ $\ldots, A_m)$ as before, but with the A_k parameters so determined that the error $\Delta=y_i-Y(x_i)$ or its square satisfies certain criteria – to be shown later.

The number of measurements

If in the interval $a=x_0<x<x_n=b$, $n+1$ interrelated (x_i, y_i) value pairs exist, and if the approximate (interpolating or equalising) function $Y(x)=Y(x, A_0, A_1, \ldots, A_m)$ is determined by m parameters, then the following cases may occur:

In interpolation the number m of parameters which determine the approximate function $Y(x)$ (with polynomials this is also the degree) must be equal to the number of measurements, i.e. $m=n+1$. This, particularly in the case of a large number of measurements, means that the approximate function will be very complicated. We draw the reader's attention to the fact that in such a case it is advisable to divide the measured range into several sections and interpolate within the sections separately, using fewer measured $(x; y)$ points. It is also advisable to draw a diagram to scale, on the basis of the measured data, which will give guidance in setting out the sections.

In equalization the number of measurements is not identical with the number of parameters which determine the approximate function. Here the following cases may occur:

— if $m>n+1$ the number of measurements must be increased because otherwise $Y(x)$ cannot be unambiguously determined:
— if $m=n+1$ determination is unambiguous, but equalization changes into interpolation;
— if $m<n+1$ there will be a large number of measured data available. This, however, will not cause any trouble because they can be made good use of in equalization, and they help also to reduce the number of data, as will be seen later.

Here is a short survey of the applications of the processes outlined.

To evaluate periodic functional relations (which can be approximated by the sum of different harmonic functions), trigonometric interpolation may be used to best advantage.

Aperiodic (monotonous or limited) functional relations can be approximated by polynomials (rational integral functions), or – less frequently – by rational fractional functions. In the former case there is a choice of whether an interpolating or an equalising polynomial should be constructed.

Interpolation with polynomials (parabolic) should be used if the measured points of function are relatively exact and subject to random errors either not at all, or to only a minimum degree. Whereas interpolating polynomials of high degree are too prone to the effects of random errors, those of low degree may leave some important pieces of information unused.

If the measured data are seriously affected by random errors, equalization is the method of choice because it smooths out, as it were, the effects of the errors and reduces the differences between their weighted mean values to a minimum, in the interval under examination.

In interpolation with a polynomial (parabolic interpolation), the function $y(x)$ tabulated by its measured points is approximated by an mth degree polynomial:

$$y(x) \cong Y(x) = A_0 + A_1 x + \ldots + A_m x^m, \tag{7.1}$$

in which $Y(x_i) = y_i$ $(i=0, 1, 2, \ldots, n)$ and, as seen earlier, $m=n+1$.

The coefficients of the approximate function (7.1) were determined by substituting all measured pairs of values (x_i, y_i) into the equation. This will yield a system of equations with m unknown quantities which, in principle, is easy to resolve but in fact needs a considerable amount of computing work.

The application of Lagrange interpolation formula facilitates parabolic interpolation. It yields the approximate function (7.1) in the following form:

$$Y(x) = L_0(x)\,y_0 + L_1(x)\,y_1 + L_2(x)\,y_2 + \ldots + L_m(x)\,y_m, \tag{7.2}$$

where

$$L_k = \frac{(x-x_0)\ldots(x-x_{k-1})(x-x_{k+1})\ldots(x-x_m)}{(x_k-x_0)\ldots(x_k-x_{k-1})(x_k-x_{k+1})\ldots(x_k-x_m)}, \tag{7.3}$$

and

$$y_k = y(x_k). \tag{7.4}$$

Here is an example of the application of the Lagrange formula to parabolic interpolation.

Example

Measuring the flow rate of an incompressible fluid through a piece of hydraulic equipment as a function of the pressure difference, yielded the following interrelated values in the 0.25...1 MPa range of pressure differences.

Pressure difference	p [MPa]	0.25	0.50	0.75	1.00
Flow rate	q [cm^3 s^{-1}]	33	46	57	66

Let us determine the interpolating polynomial $q(\Delta p)$ in this interval. In what follows we shall not include the units because they will not be needed. In accordance with our previous notations of y and x let $\Delta p=x$ and $q=y$. This changes the above table to:

x	$x_0=0.25$	$x_1=0.50$	$x_2=0.75$	$x_3=1.00$
y	$y_0=33$	$y_1=46$	$y_2=57$	$y_3=66$

The Lagrange polynomials as in (7.3):

$$L_0 = \frac{(x-x_1)(x-x_2)(x-x_3)}{(x_0-x_1)(x_0-x_2)(x_0-x_3)} = \frac{(x-0.5)(x-0.75)(x-1)}{(0.25-0.5)(0.25-0.75)(0.25-1)} =$$

$$= \frac{x^3-2.25x^2+1.625x-0.375}{-0.093\,75};$$

$$L_1 = \frac{(x-x_0)(x-x_2)(x-x_3)}{(x_1-x_0)(x_1-x_2)(x_1-x_3)} = \frac{(x-0.25)(x-0.75)(x-1)}{(0.5-0.25)(0.5-0.75)(0.5-1)} =$$

$$= \frac{x^3-2x^2+1.187\,5x-0.187\,5}{+0.0312\,5};$$

$$L_2 = \frac{(x-x_0)(x-x_1)(x-x_2)}{(x_2-x_0)(x_2-x_1)(x_2-x_3)} = \frac{(x-0.25)(x-0.5)(x-0.75)}{(0.75-0.25)(0.75-0.5)(0.75-1)} =$$

$$= \frac{x^3-1.75x^2+0.875x-0.125}{-0.031\,25};$$

$$L_3 = \frac{(x-x_0)(x-x_1)(x-x_2)}{(x_3-x_0)(x_3-x_1)(x_3-x_2)} = \frac{(x-0.25)(x-0.5)(x-0.75)}{(1-0.25)(1-0.5)(1-0.75)} =$$

$$= \frac{x^3-1.5x^2+0.687\,5x-0.093\,75}{+0.093\,75}.$$

Given the Lagange polynomials, the interpolation polynomial for our case assumes the following form:

$$Y(x) = L_0(x)\,y_0+L_1(x)\,y_1+L_2(x)\,y_2+L_3(x)\,y_3.$$

Substituting the Lagrange polynomials determined before: $L_i(x)$ $(i=0...3)$ and the values of $y_0 ... y_3$ from the table (measurements) and by the reduction of the coefficients of equal power (without showing the simple but extensive calculations), we find that the coefficient of x^3 is zero, and the interpolating polynomial

$$Y(x) = -16x^2+64x+17.9.$$

Note: The elimination of the coefficient A_3 of the interpolating polynomial whereby it becomes of second degree, has real physical significance as well. Namely, with an incompressible medium the throughflow characteristic appears as a second degree parabola (modified Bernoulli equation), and to describe it, there is no need for a third degree member.

7.1.1.2 TRIGONOMETRIC INTERPOLATION

In trigonometric interpolation the function $y(x)$, tabulated by the points of the graph is approximated by a trigonometric polynomial consisting of harmonic members.

Assume that measurements yielded a number n of functional values $y(x_k)=y_k$ corresponding to the independent variables $\left(x_k=\frac{kT}{n}(k=0, 1, 2, \ldots, n-1)\right)$, i.e., a number n of related pairs of values $(x_k, y_k.)$

In trigonometric interpolation, the function $y(x)$ in the $(0, T)$ interval, is approximated by the trigonometric polynomial

$$y(x) \cong Y(x) = \frac{A_0}{2} + \sum_{i=1}^{n}\left(A_i \cos i\frac{2\pi x}{T} + B_i \sin i\frac{2\pi x}{T}\right) \tag{7.5}$$

in such a way as to reduce the square mean error $\sum_{k=0}^{n-1}[Y(x_k)-y_k]^2$ to a minimum. (Remember that $m<n/2$.)

The corresponding coefficients A_i and B_i are

$$A_i = \frac{2}{n}\sum_{k=0}^{n-1} y_k \cos i\frac{2\pi k}{n} \tag{7.6a}$$

$$B_i = \frac{2}{n}\sum_{k=0}^{n-1} y_k \sin i\frac{2\pi k}{n} \tag{7.6b}$$

$$\left(0 \leqq i \leqq \frac{n}{2}\right).$$

If $m=n/2$ then from the relationship $A_m=A_{n/2}=\frac{1}{n}\sum_{k=0}^{n-1}(-1)^k y_k$, $Y(x_k)=y(x_k)$ will apply to arbitrary B_m, which means that at the given points the approximate function yields accurate values. Although trigonometric interpolation can be applied to any $f(x)$, it is applied mostly to periodic functions.

The sums (7.6a) and (7.6b) can be calculated with ease, provided that n is divisible by 4. Runge [20], for the case of $n=12$

$$y_k = y\left(\frac{kT}{12}\right),$$

compiled Table 7.1 which provides instructions for the work of calculation.

The case is particularly simple if we wish to fit a trigonometric polynomial which has no harmonics above the third to a functional value determined by 12 measurements. In this case there is no need to compile a table, and the amplitude of the harmonic components can be calculated on the basis of simpler numerical formulae:

TABLE 7.1.

12-ordinate Runge arrangement for trigonometric interpolation

Row	Obtained by							
1	Given (measured) functional values	y_0	y_1	y_2	y_3	y_4	y_5	y_6
2			y_{11}	y_{10}	y_9	y_8	y_7	
3	The sum of 1 and 2	s_0	s_1	s_2	s_3	s_4	s_5	s_6
4	The difference between 1 and 2		d_1	d_2	d_3	d_4	d_5	
5	Copying	s_0	s_1	s_2	s_3	d_1	d_2	d_3
6		s_6	s_5	s_4		d_5	d_4	
7	The sum of 5 and 6;	s_0	s_1'	s_2'	s_3'	s_1''	s_2''	s_3''
8	The difference between 5 and 6	d_0'	d_1'	d_2'		d_1''	d_2''	

Row	Obtained by														
9	Copying and multiplication	s_0'	s_1'	d_0'		s_0	$-s_3'$	d_0'	d_2'	s_3''				s_1''	s_3''
10		s_2'	s_1'												
11				$1/2d_2'$		$-1/2s_0'$	$1/2s_2'$			$1/2s_1''$					
12					$\sqrt{3}/2d_1'$						$\sqrt{3}/2s_2''$	$\sqrt{3}/2d_1''$	$\sqrt{3}/2d_2''$		
13	The columns are to be added (9–12)	S_1	T_1	S_2	T_2	S_3	T_3	S_4	T_4	S_5	T_5	S_6	T_6	S_7	T_7
14	S_i+T_i	$6A_0$		$6A_i$		$6A_2$				$6B_1$		$6B_2$			
15	S_i-T_i	$12A_6$		$6A_5$		$6A_4$		$6A_3$		$6B_5$		$6B_4$		$6B_3$	

$$
\left.\begin{aligned}
6A_0 &= y_0+y_1+y_2+y\ldots+y_{11} && a\\
A_1 &= \frac{1}{2}(y_0-y_6)+A_3 && b\\
4A_2 &= y_0-y_3+y_6-y_9 && c\\
6A_3 &= y_3-y_2+y_4-y_6+y_8-y_{10} && d\\
4B_1 &= \frac{1}{2}(y_3-y_9)+B_3 && e\\
4B_2 &= y(T/8)-y(3T/8)+y(5T/8)-y(7T/8) && f\\
6B_3 &= y_1-y_3+y_5-y_7+y_9-y_{11}. && g
\end{aligned}\right\} \tag{7.7}
$$

Note that the four functional values required for (7.7) can be, or are, measured from the scale graph previously plotted on the basis of the measured data of the periodic function to be interpolated with the trigonometric polynomial.

Since this ordinate arrangement may also be applied to recordings made in practice, this process is illustrated by an example:

Example

The recording shown in Fig. 7.1 was plotted from measurements. In the figure, the recording shows the following:

$$
\begin{aligned}
&y_0 = 1;\ y_1 = 0.8;\ y_2 = 0.5;\ y_3 = 0;\ y_4 = -0.9;\ y_5 = -1.5;\\
&y_6 = -1.5;\ y_7 = -0.9;\ y_8 = 0;\ y_9 = 0.5;\ y_{10} = 0.8;\ y_{11} = 1,\\
&T = 11;\ T/8 = 1.375;\ 3T/8 = 4.125;\ 5T/8 = 6.875,\\
&7T/8 = 9.625;\ y(T/8) = 0.8;\ y(3T/8) = -1;\ y(5T/8) = -1,\\
&y(7T/8) = 0.8.
\end{aligned}
$$

Numerically the Eqs. (7.7) are as follows:

$$
\begin{aligned}
6A_0 &= 1+0.8+0.5+0-0.9-1.5-1.5-0.9+0+0.5+0.8+1 = -0.2,\\
A_1 &= 1/2[1-(-1{,}5)]+A_3 = 1.25-0.1166 = 1.133,\\
4A_2 &= 1-0-1.5-0.5 = -1,\\
6A_3 &= -0.5-0.9+1.5-0.8 = -0.7,\\
B_1 &= 1/2\times 0.5-0.5 = -0.25,\\
4B_2 &= 0.8-(-1)+(-1)-0.8 = 0,\\
6B_3 &= 0.8-0+(-1.5)-(-0.9)+0.5-1 = -0.3.
\end{aligned}
$$

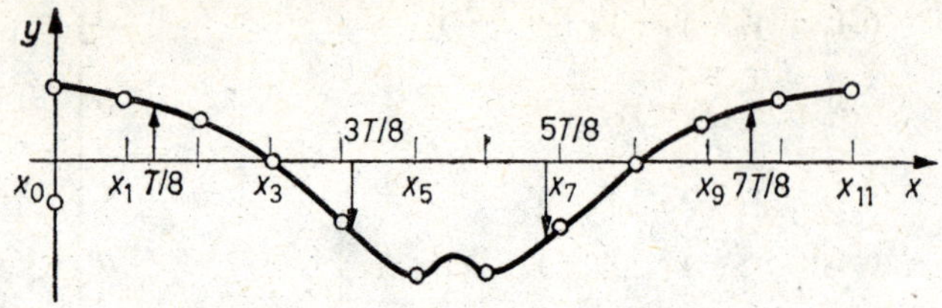

Fig. 7.1. *Recording for the numerical example illustrating the process of trigonometric interpolation*

Solving the equations yields the values:

$$A_0/2 = -0.0166,$$
$$A_1 = 1.133,$$
$$A_2 = -0.25,$$
$$A_3 = -0.1166,$$
$$B_1 = -0.25,$$
$$B_2 = 0,$$
$$B_3 = -0.3,$$

which yield the approximate trigonometric polynomial that analytically describes the recording as:

$$Y(x) = -0.0166 + 1.133 \cos\frac{2\pi}{11}x - 0.25 \cos\frac{4\pi}{11}x -$$

$$-0.0166 \cos\frac{6\pi}{11}x - 0.25 \sin\frac{2\pi}{11}x - 0.3 \sin\frac{6\pi}{11}x.$$

7.1.1.3 EQUALIZATION

Let us once more examine the function $y = y(x)$ by measurement. Let us assume that there are $n+1$ discrete measured values $y_i = y(x_i)$, $(i = 0, 1, 2, \ldots, n)$. The approximate function $Y(x)$ is:

$$y(x) \cong Y(x) = Y(x, A_0, A_1, A_2, \ldots, A_m) \qquad (7.8)$$

such that

$$\Delta_i = y_i - Y(x_i) \qquad (7.9)$$

the error in the sum of the errors Δ_i defined at the ith point by the equation satisfies certain criteria. We shall deal with three cases, of which two are of considerable significance in practical measurement evaluation. They are

— the method of equal rate of approximation,
— the method of mean values,
— the method of least squares.

In the case of approximation at an equal rate, the sum of the absolute errors, considering all points in the set of discrete points determined by measurement should be

$$\sum_{i=0}^{n} |\Delta_i|_{max} = \text{minimum!} \tag{7.10}$$

i.e. minimal. Although in theory this criterion offers the best approximation, if it is met the calculation of the coefficients (7.8) will require a rather laborious iteration. In the literature this method of approximation is called the Chebyshev approximation, after its inventor. Because of its sophistication it is seldom used in numerical calculation, but is used primarily in theoretical work.

The method of mean values

The method of mean values is simple and easy to use. Let us assume once more that in the interval $a=x_0<x<x_n=b$ our measurements yielded $(n+1)$ related pairs of values (x_i, y_i). Let us assume also that the approximate function given in (7.8) is a polynomial of the mth degree:

$$y(x) \cong Y(x) = A_0+A_1x+A_2x^2+\ldots+A_mx^m, \tag{7.11}$$

which in form is equivalent to (7.1) but $m>n+1$. In interpolation we made the substitution $Y(x_i)=y_i$: our aim now is to compensate the errors in the measurement. We may assume that the signs and magnitude of the errors vary with equal probability and that they thus compensate for one another, and yield zero as their mean value. Accordingly, with the method of mean values:

$$\sum_{i=1}^{n} \Delta_i = 0. \tag{7.12}$$

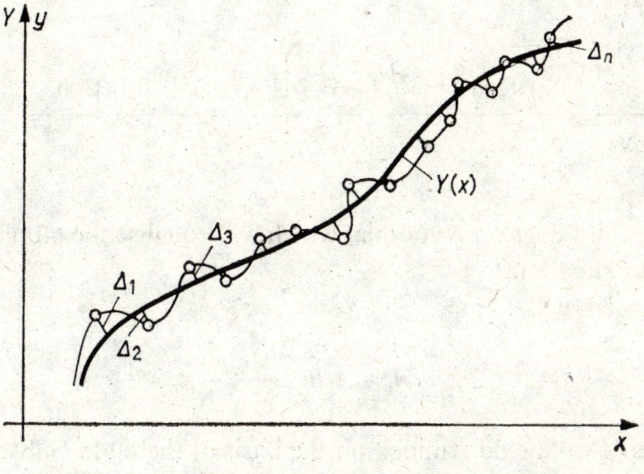

Fig. 7.2. *Equalization with the method of mean values*

So, if the measured points are given (Fig. 7.2) we may then derive a (probable) approximate function $Y(x)$ which satisfies (7.12). Since it has been assumed that (x_i, y_i) contain errors, the error defined by (7.9) is substituted into (7.11). Substituting at the ith measured point and rearranging the equation for Δ_i we have

$$\Delta_i = A_0+A_1x+A_2x^2+\ldots+A_mx^m-y_i. \tag{7.13}$$

If we substitute our measured points into this equation, we obtain a system of $(n+1)$ equations with $m+1$ unknowns.

$$\left.\begin{aligned} \Delta_0 &= A_0+A_1x_0+A_2x_0^2+\ldots+A_mx_0^m-y_0 \\ \Delta_1 &= A_0+A_1x_1+A_2x_1^2+\ldots+A_mx_1^m-y_1 \\ &\ldots\ldots\ldots\ldots\ldots\ldots\ldots\ldots\ldots\ldots \\ \Delta_n &= A_0+A_1x_n+A_2x_n^2+\ldots+A_mx_n^m-y_n \end{aligned}\right\} \tag{7.14}$$

Since $m>n+1$ the Eqs (7.14) must be split into as many groups as there are parameters A_i to be determined. The criterion $\sum_i \Delta_i=0$ will also apply to these groups. The only shortcoming in the method lies in the fact that the value of m must be arbitrarily chosen, and this choice needs care and some experience.

Here is a simple numerical example to illustrate the practical application.

Example

The thermovoltage vs. temperature characteristics of a thermoelement W−WRe$_{30}$ $U_T(\vartheta)$ were established by measurement. The results were:

ϑ (°C)	0	200	400	600	800	1000
U_T (mV)	0	2	4.5	7	10.5	15.5

ϑ (°C)	1200	1400	1600	1800	2000
U_T (mV)	21.3	26.1	30.4	34	39.2

Now find the third degree polynomial which will equalise the tabulated function by equal rate approximation.

That polynomial will be:

$$U_T = A_0+A_1\vartheta+A_2\vartheta^2+A_3\vartheta^3$$

and its coefficients will be determined on the basis of the table. Substituting at the 11 measured points we obtain 11 equations:

$$0 = A_0+A_1\cdot 0+A_2\cdot 0+A_3\cdot 0 \qquad (A_0 = 0 \text{ follows})$$

$$\left.\begin{aligned} 2 &= 2\cdot 10^2 A_1+4\cdot 10^4 A_2+8\cdot 10^6 A_3\\ 4.5 &= 4\cdot 10^2 A_1+1.6\cdot 10^5 A_2+6.4\cdot 10^7 A_3\\ 7 &= 6\cdot 10^2 A_1+3.6\cdot 10^5 A_2+2.16\cdot 10^8 A_3 \end{aligned}\right\}\ \text{I}$$

$$\left.\begin{aligned} 10.5 &= 8\cdot 10^2 A_1+6.4\cdot 10^5 A_2+5.12\cdot 10^8 A_3\\ 15.5 &= 1\cdot 10^3 A_1+1\cdot 10^6 A_2+1\cdot 10^9 A_3\\ 21.3 &= 1.2\cdot 10^3 A_1+1.44\cdot 10^6 A_2+1.728\cdot 10^9 A_3\\ 26.1 &= 1.4\cdot 10^3 A+1.96\cdot 10^6 A_2+2.744\cdot 10^9 A_3 \end{aligned}\right\}\ \text{II}$$

$$\left.\begin{aligned} 30.4 &= 1.6\cdot 10^3 A_1+2.56\cdot 10^6 A_2+4.096\cdot 10^9 A_3\\ 34 &= 1.8\cdot 10^3 A_1+3.24\cdot 10^6 A_2+5.832\cdot 10^9 A_3\\ 39.2 &= 2\cdot 10^3 A_1+4\cdot 10^6 A_2+8\cdot 10^9 A_3. \end{aligned}\right\}\ \text{III}$$

From the first equation, $A_0=0$. The remaining 10 equations seem to carry more information than is necessary for the determination of the remaining three unknowns, but there is the possibility of reducing the amount of data by combining the equations and reducing their number. To determine the remaining three unknowns A_1, A_2 and A_3 only three equations will be required.

Adding e.g. the first three, the second four and the third three equations, the original 11 equations can be reduced to three, which carry all the information:

$$13.5 = 12\times 10^2 A_1+56\times 10^4 A_2+288\times 10^6 A_3,$$

$$73.4 = 4.4\times 10^3 A_1+5.04 A_2+5.984\times 10^9 A_3,$$

$$103.6 = 5.4\times 10^3 A_1+9.8\times 10^6 A_2+17.928\times 10^9 A_3.$$

The set of equations with three unknowns A_1, A_2 and A_3 can also be solved by the determinant method. Omitting the detailed calculation, for the three unknowns we have the values

$$A_1 = 5.673\times 10^{-3},$$

$$A_2 = 1.375\times 10^{-5},$$

$$A_3 = -3.887\times 10^{-9},$$

so

$$U_T = 5.673\times 10^{-3}\vartheta+1.375\times 10^{-5}\vartheta^2-3.887\times 10^{-9}\vartheta^3.$$

It might be instructive to plot the measured points and the points yielded by the resulting polynomials using the same coordinates.

The method of least squares requires that the square of the error in Δ_i defined by Eq. (7.9) should be minimal.

$$\Delta_i^2 = \sum_{i=0}^{n} [y_i-Y(x_i)]^2 = \text{minimum}. \tag{7.15}$$

TABLE 7.2.

Summary of the more important approximation formulae

Type and mathematical form of the formula	Linearization relations and remarks	The curves
1. Power function $y = ax^b$	$\log y = \log a + b \cdot \log x$ $Y = \log y;\ X = \log x$ $\boxed{Y = \log a + bX}$ or $Y = y;\ X = x^b$ $\boxed{Y = aX}$ *a–c:* integral exponent; *d–j:* fractional exponent; the figures show the effect of the exponent	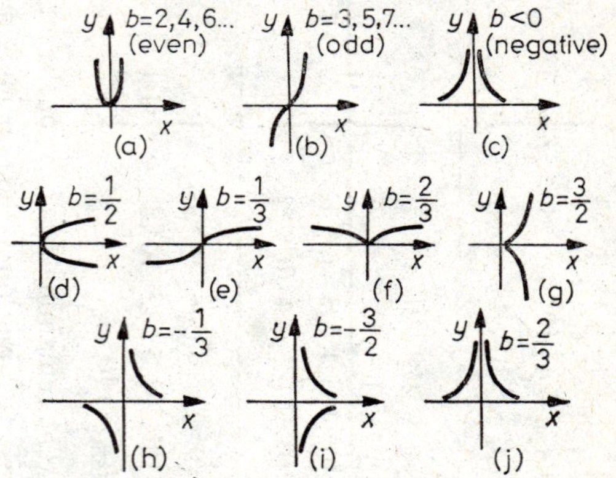
2. Power function with additive constant $y = ax^b + b$	$Y = y;\ X = x^b$ $Y = aX + c$	The curves are the same as in 1, but shifted along the y-axis by c; corresponding to $1a$ — to $1d$ 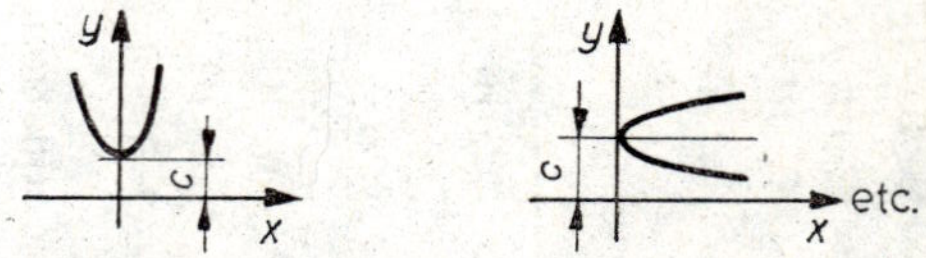
3. Exponential function $y = ae^{bx}$ ill. $y = ab^{cx}$	$\log y = \log a + xb \log e$ $Y = \log y;\ X = x$ $\boxed{Y = \log a + (b \cdot \log e)X}$	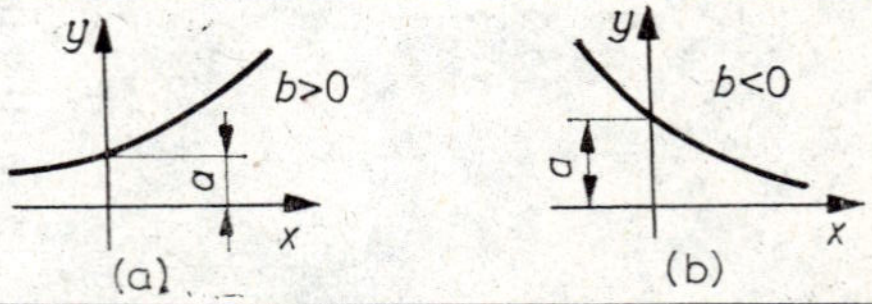

9*

<table>
<tr><td>4. Exponential function with additive constant

$y = ae^{bx}+c$</td><td>$\log y-c = \log a+xb\cdot\log e$
$Y = \log y-c;\ X = x$
$\boxed{Y = \log a+(b\cdot\log e)X}$</td><td>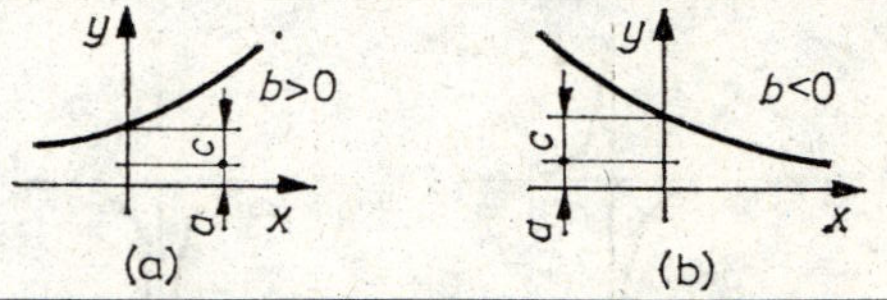</td></tr>
<tr><td>5. Density function of normal probability distribution and its form shifted along the x-axis

(a) $y = Ke^{-(ax)^2}$
(b) $y = Ke^{-a^2(x-b)^2}$</td><td>(a) $\ln y = \ln K-(ax)^2\ln e$
$Y = \ln y;\ X = x^2$
$\boxed{Y = \ln K-a^2\ln e\cdot X}$
(b) $\ln y = \ln K-a^2x-b^2\ln e$
$Y = \ln y;\ X = x-b^2$
$\boxed{Y = \ln K-a^2\ln e\cdot X}$</td><td>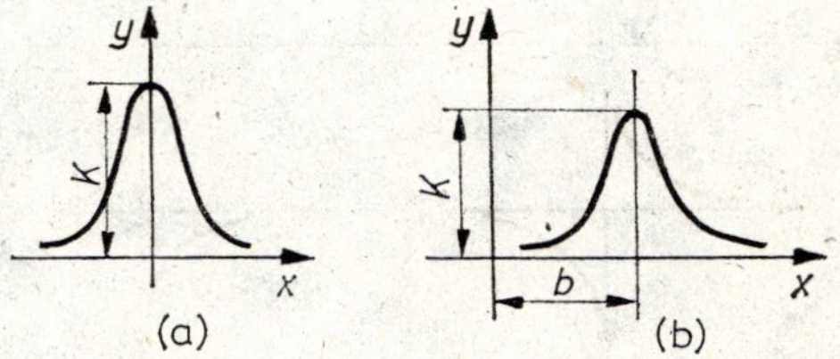</td></tr>
<tr><td>6. Parabola usually used for technical interpolations and/or approximations

$y = ax^2+bx+c$</td><td>The linearisation relations can be determined in the following way: choosing an arbitrary point on the curve whose co-ordinates are (x_iy_i), so $y_1 = ax_1^2+bx_1+c$, deducting it from the basic equation and after simple transformations obtaining the equation from which
$\frac{y-y_1}{x-x_1} = a(x+x_1)+b$
$Y = \frac{y-y_1}{x-x_1};\ X = x$
thus
$\boxed{Y = aX+ax_1+b}$</td><td>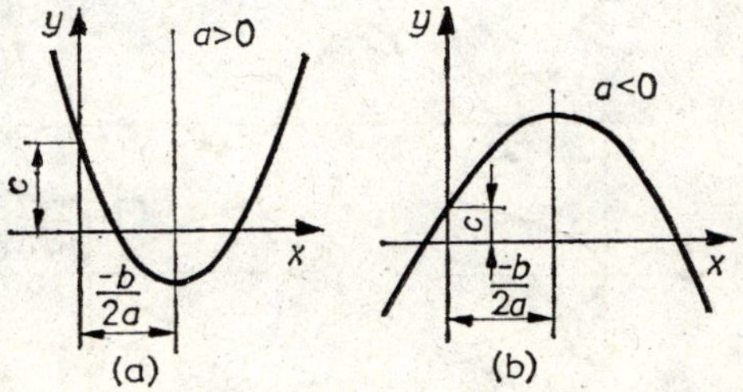</td></tr>
</table>

TABLE 7.2. *(cont.)*

Type and mathematical form of the formula	Linearization relations and remarks	The curves
7. Square root of the linear function $y^2 = ax+b$ $y = \pm\sqrt{ax+b}$	$Y = y^2;\ X = x$ $Y = aX+b$	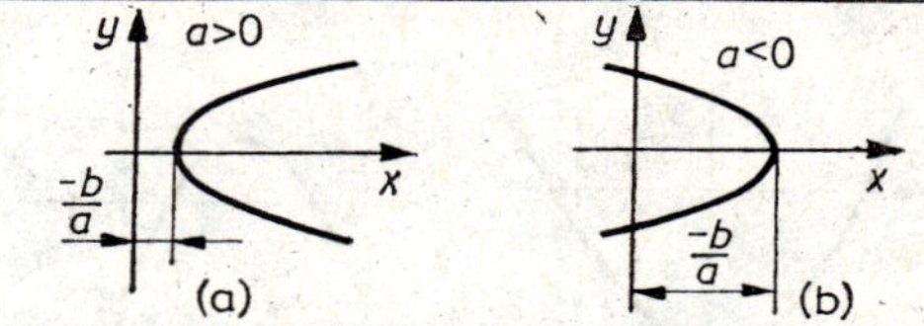
8. Square root of the function of second degree $y^2 = ax^2+bx+c$ $y = \pm\sqrt{ax^2+bx+c}$	Introducing the auxiliary variable $y = y^2$ the case can be reduced to 6. $Y = \frac{y-x_1}{x-x_1} = \frac{y^2-y_1^2}{x-x_1};\ X = x$ and thus $Y = aX+ax_1+b$ $\Delta = 4ac-b$	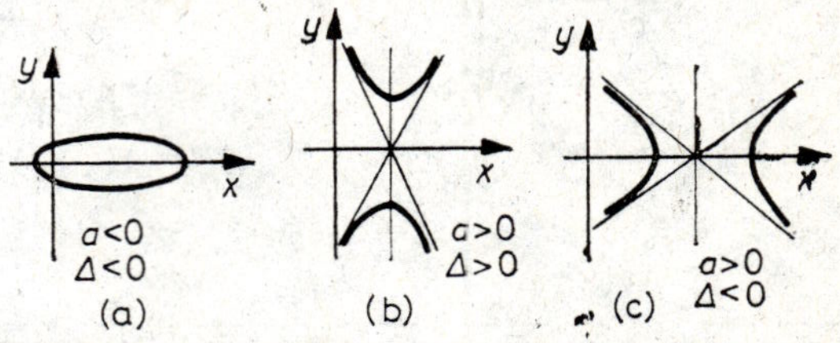
9. Inverse proportionality $y = \frac{a}{x}$	$Y = y;\ X = \frac{1}{x}$ $\boxed{Y = aX}$ Note: The functional curve is a symmetrical hyperbola with the x- and y-axes as its asymptotes	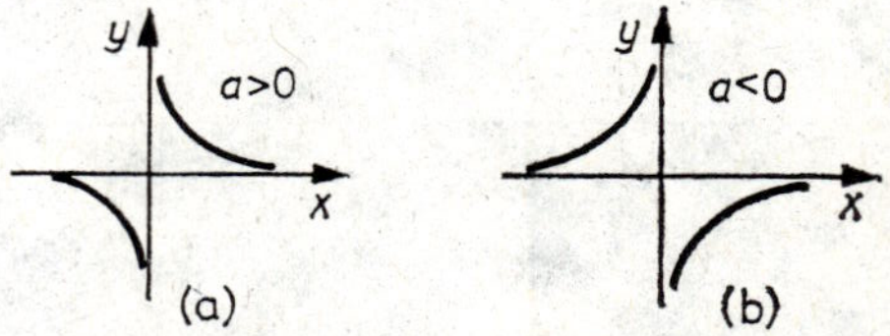
10. Polytropic (adiabatic) change of state described by function $xy^b = a$ i.e. $y = \frac{a}{x^b} = ax^{-b}$	$\ln y = \ln a - b \ln x$ $Y = \ln y;\ X = \ln x$ $\boxed{Y = \ln a - bX}$ Note: a hyperbolic (but not symmetrical) function	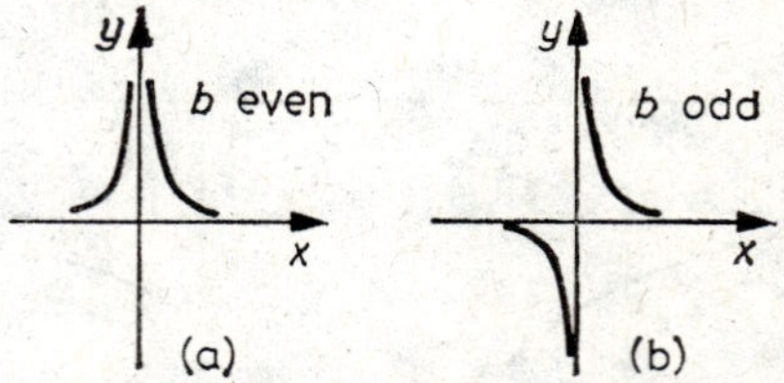

11. Linear fractional function $y = \dfrac{ax+b}{cx+d}$	As in case 6 with an arbitrary $x_1 y_1$ point selected on the functional curve, $y_1 = \dfrac{ax_1+b}{cx_1+d}$ deducing it from the basic equation and after a simple transformation we have the equation $\dfrac{x-x_1}{y-y_1} = \dfrac{cx_1+d}{da+bc}(cx+d)$ whence $Y = \dfrac{x-x_1}{y-y_1};\; X = x$ and thus $\boxed{Y = \dfrac{cx_1+d}{da+bc}(cX+d)}$ Note: Symmetrical hyperbola but its asymptotes are not the x and y axes!	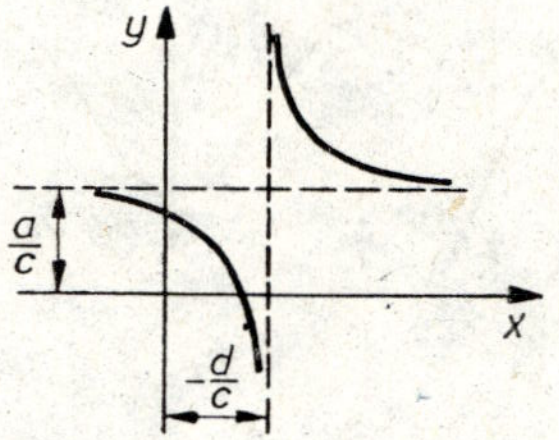
12. Third degree curve $y = \dfrac{1}{ax^2+bx+c}$	Introducing the auxiliary variable $\bar{y} = \dfrac{1}{y}$ the case can be reduced to 6. $Y = \dfrac{\bar{y}-\bar{y}_1}{x-x_1} = \dfrac{\frac{1}{y}-\frac{1}{y_1}}{x-x_1};\; X = x$ and thus $\boxed{Y = aX+ax_1+b}$ Notation: $\Delta = 4ac-b^2$	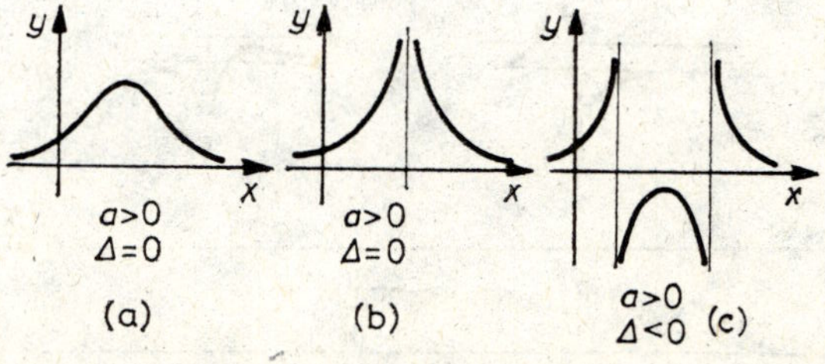

TABLE 7.2. (*cont.*)

Type and mathematical form of the formula	Linearisation relations and remarks	The curves
13. Third degree curve $y = b+\frac{b}{x}+\frac{c}{x^2}$ $b \neq 0$ $c \neq 0$	Introducing the auxiliary variable $\bar{x} = \frac{1}{x}$ the case can be reduced to 6. $\log y = \log a+bx\cdot\log e+cx^2\cdot\log e$ $Y = \frac{y-y_1}{\bar{x}-\bar{x}_1} = \frac{y-y_1}{\frac{1}{x}-\frac{1}{x_1}}; \ X = \bar{x} = \frac{1}{x}$ and thus $\boxed{Y = aX+ax_1+b}$	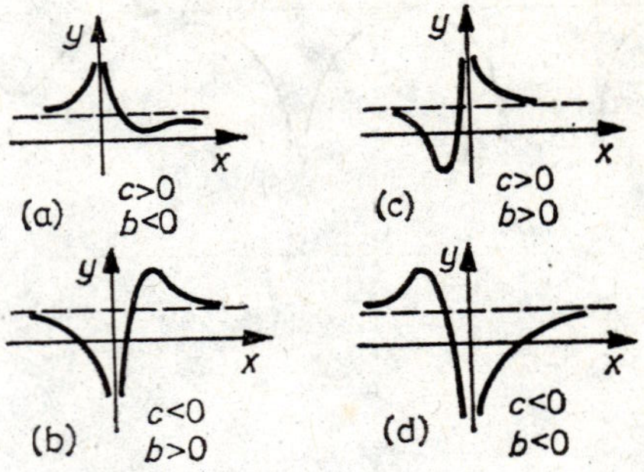
14. Exponential function with rational integral functional exponent $y = ae^{bx+cx^2}$	Introducing the auxiliary variable $\bar{y} = \log y$ the case can be reduced to 6. $\log y = \log a+bx\cdot\log e+cx^2 \log e$ $Y = \frac{y-y_1}{x-x_1} = \frac{\lg\frac{y}{y_1}}{x-x_1}; \ X = x$ and thus $\boxed{Y = c\cdot\log eX+c\cdot\log ex_1+b\cdot\log e}$ Note: If $a<0$, the mirror image of the curves in the x-axis will apply.	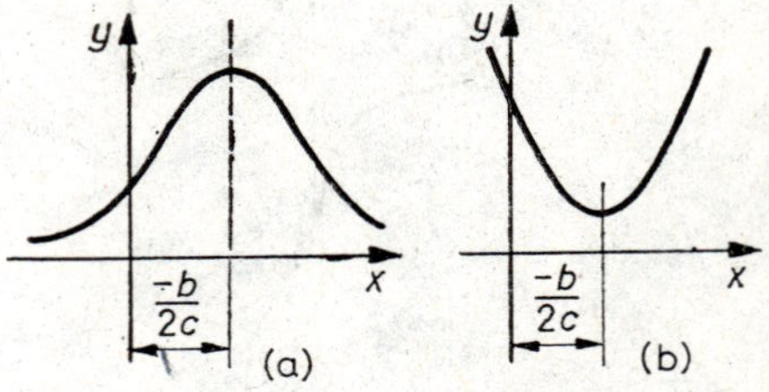

When this method is applied, the approximate function $Y(x)$ is sought in the following form:

$$y(x) \cong Y(x) = A_0 f_0(x) + A_1 f_1(x) + A_2 f_2(x) + \ldots + A_m f_m(x). \qquad (7.16)$$

To simplify the calculation the functions $f_i(x)$ should be ith degree polynomials, which according to

$$\sum_{i=0}^{n} f_j(x_i) f_k(x_i) = 0 \quad (j \neq k) \qquad (7.17)$$

are orthogonal.

For the sake of completeness, the reader should remember that such polynomials can be obtained from the expressions $1, x, x^2, \ldots$ by orthogonalization according to Gram—Schmidt [20]. As the end result, the coefficients of A_j, which in equation (7.16) give the minimal square error, can be derived from the following relationship:

$$A_j = \frac{\sum_{i=0}^{n} y(x_i) f_j(x_i)}{\sum_{i=0}^{n} f_j^2(x_i)} \quad (j = 0, 1, 2, \ldots, m \leqq n). \qquad (7.18)$$

If $m=n$, the polynomial so obtained will be an interpolating polynomial. If $m<n$, even the addition of a new member of $A_i f_j(x)$ will not affect the previous members.

7.1.2 CHOOSING THE BEST APPROXIMATING FORMULA

The (empirical) functional relation $y=y(x)$ determined by measurement can also be interpreted mathematically by selecting a function of known form, whose shape is similar to that of $y(x)$ and whose parameters have been determined from the data related to $y(x)$.

This process, however, can only then be continued if no physical considerations affect the formula to be chosen, because in other cases the physical considerations are overriding.

In practice we may choose a function which is easy to construct and then establish by linearization whether it is really suitable for the approximation to our measured function. This is important because the comparison of plots merely by eye may lead to false conclusions.

The essential task in linearization is to seek for the function $y=y(x)$ two suitable functions $X=\varphi(x, y)$ and $Y=\psi(x, y)$ for which the criterion $Y=Y(X)$ is linear. In practice this is quite easy. With the values of x_i and y_i determined by measurement, and from the functions φ and ψ, it is possible to determine whether the measured functional relation and the chosen type of formula are close to one another, i.e. whether the chosen formula is suitable for approximating $y(x)$.

In Table 7.2 we have compiled some of the more important types of formula, their linear relations, and the curves they yield. The application of the method will be illustrated by an example.

Example

The first two rows of Table 7.3 give the functional relation $y(x)$; Fig. 7.3 is plotted from the table to scale. Using the figure, we make a comparison with the curves of Table 7.2 and find that the cases 5b and 12 show a striking resemblance. In 5b:$Y=\ln y$ and $X=x-b^2$ while in 12: $Y=\dfrac{1/y-1/y_1}{x-x_1}$ and $X=x$ are the linearization relations.

TABLE 7.3.

For the numerical example of the selection of the best approximation formula

Name		Numerical data					
Measured value	x	0	0.2	0.4	0.6	0.8	1.0
	y	0.36	0.52	0.69	0.85	0.96	1.0
Table 7.2. Fig. 5b	X	1.0	0.64	0.36	0.16	0.4	0
	Y	−1.0	−0.64	−0.36	−0.16	−0.4	0
Table 7.2. Fig. 12.	X	0	0.2	0.4	0.6	0.8	1.0
	Y	−1.71	−1.12	−0.72	−0.43	−0.20	0
Measured value	x	1.2	1.4	1.6	1.8	2.0	
	y	0.96	0.85	0.69	0.52	0.36	
Table 7.2. Fig, 5b	X	0.04	0.16	0.36	0.64	1.0	
	Y	−0.04	−0.16	−0.36	−0.64	−1.0	
Table 7.2. Fig. 12.	X	1.2	1.4	1.6	1.8	2.0	
	Y	0.20	0.43	0.72	1.12	1.71	

Note: In 5b of Table 7.2 $X=x-b^2$, $Y=\ln y$

In 12 of Table 7.2: $X=x$, $Y=\dfrac{1/y-1/y_1}{x-x_1}$

The rest of Table 7.3 was obtained from the linearization relations, and calculating the values of Y and X. From this we plotted Figs 7.4a and b which prove that $y=Ke^{-a^2(x-b)^2}$, corresponding to 5b, can be applied to advantage. Selecting three

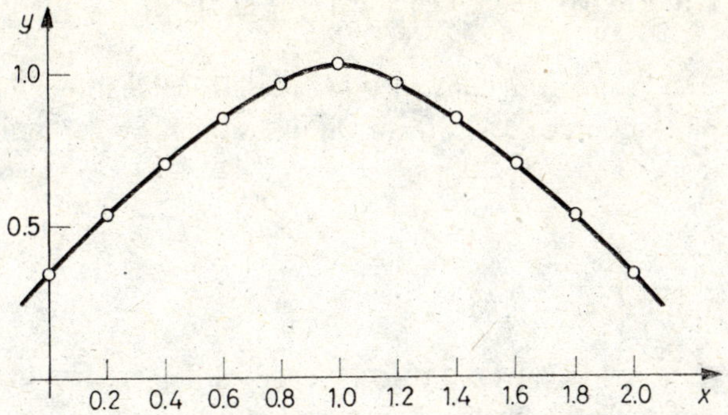

Fig. 7.3. *Guide for the selection of the best approximation formula*

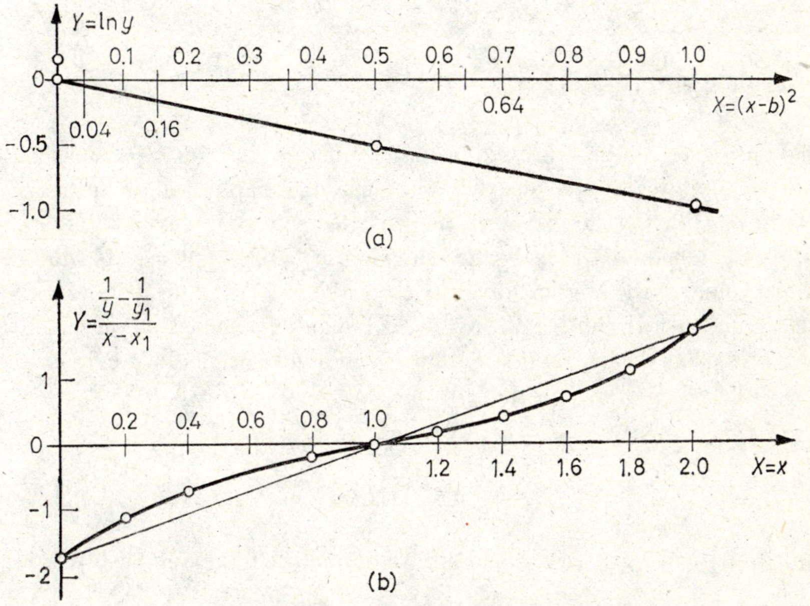

Fig. 7.4. (a) *Easily linearised,* (b) *less easily linearised formula*

characteristic points, e.g. $x=0$, 1, and 2, we may write the following equations:

$$\ln 0.36 = \ln K + a^2 b^2,$$

$$\ln 0.36 = \ln K - a^2(2-b)^2 \quad \text{and}$$

$$\ln 1 = \ln K - a^2(1-b)^2.$$

From the three equations the three unknowns K, a, and b can be determined. By calculation (not shown here), we find that $K=1$, $a=1$ and $b=1$.

7.2 GRAPHICAL METHODS [7.5], [7.6], [7.7]

In addition to the geometric construction, (i.e., Euclidean geometry) graphical methods are all plotting processes, which, within estimable error limits, yield useful results.

7.2.1 BASIC CONCEPTS

In graphical methods physical (and/or mathematical) quantities are illustrated or interpreted by unambiguously matched measurable quantities (distance, angle, area, vectors, etc.) and the mathematical operations to be executed on them are replaced by diagrams, constructions and the use of mathematical instruments; the results are then measured or read out on a scale which corresponds to the pre-selected units.

7.2.1.1 UNITS, SCALE RATIOS

Physical quantities can be expressed as the product of their numbers and their units. Since, in what follows, we shall deal with the illustration of physical quantities by drawings, we must differentiate between the so-called physical unit f_x of the physical quantity x examined, and the so-called drawing scale of the plot l_x. Let the number (numerical value) of the physical quantity x be $[x]$ and let X be the distance (in mm or any chosen unit) which illustrates the physical quantity x. Since, as we said, x and X can be expressed as the product of a number and a unit, using our previous notation, we may write

$$x = [x] f_x, \tag{7.19}$$

$$X = [x] \, l_x. \tag{7.20}$$

Eliminating $[x]$ in Eqs (7.19) and (7.20) we obtain the following basic relationship:

$$\frac{x}{X} = \frac{f_x}{l_x}. \tag{7.21}$$

Example

$p = 5\ \mathrm{N/m^{-2}} \equiv 5\ \mathrm{Pa}$ (let us denote pressure by x, so $x = 5$ Pa). Let us represent it in the drawing by 20 mm denoted by

$$x = 5\ \mathrm{Pa} \triangleq X = 20\ \mathrm{mm}.$$

With our notation:

$$x = 5, \quad f_x = \mathrm{Pa}, \quad X = 20\ \mathrm{mm}.$$

Let us find the scale which, using (7.21), gives

$$l_x = \frac{f_x X}{x} = \frac{\text{Pa} \times 20\ \text{mm}}{5\ \text{Pa}} = 4\ \text{mm}.$$

This gives for the drawing 1 Pa $\triangleq$ 4 mm.

Selecting the scale

If in a diagram where $[X]$ is to be plotted, the length of X is given (e.g. by the size of the sheet) then l_x should be chosen on the basis of the following relationship:

$$l_x \leqq \frac{X}{x}. \tag{7.20a}$$

(i.e. l_x should be rounded down). Another requirement in choosing the scale is that of accuracy. In drawings the average error in measurement or drafting is approximately 0.15 mm, equal to $0.15/l_x$ dimensional units. If the degree of accuracy is specified, it is possible to determine the required scale, i.e. the minimal size of the drawing. The errors in areas can be calculated by squaring. Similar considerations apply if angles are measured. If the angle will be determined in terms of distances, then the error in its tangent must be evaluated.

When plotting measured functional relations, it is important to choose the scales l_x and l_y for the curve $y(x)$ in such a way that they yield a plot which is neither too flat nor too steep. This will ensure satisfactory accuracy. The accuracy of graphical processes is satisfactory if the slope of the curve drawn to scale is between 25 and 75°.

For simplicity's sake, it is preferable to assume a monotonous (continuously decreasing or increasing) functional curve (Fig. 7.5). Measurements were executed in the interval (x_0, x_n). Since:

$$X = l_x[x], \tag{7.22a}$$

$$Y = l_y[y], \tag{7.22b}$$

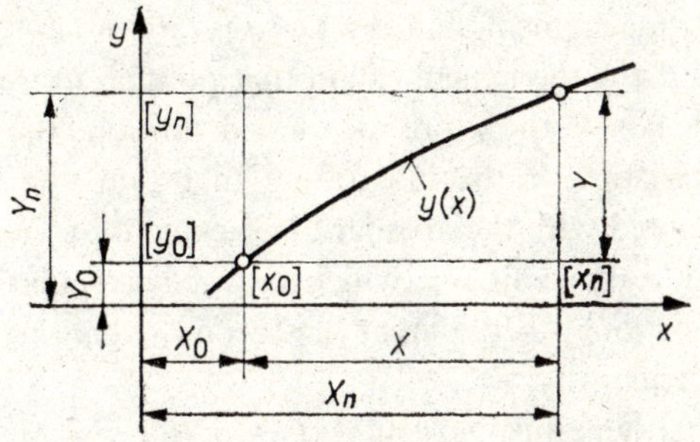

Fig. 7.5. *Guidance for the selection of scales*

on the basis of Fig. 7.5 we may write that

$$\tan\alpha = \frac{Y_n - Y_0}{X_n - X_0} = \frac{Y}{X} = \frac{l_y([y_n]-[y_0])}{l_x([x_n]-[x_0])} = \frac{l_y[y]}{l_x[x]}. \tag{7.23}$$

If the x- and y-axes are drawn to the same scale then $\tan\alpha$ can be calculated directly from the numerical values. With the recommended slope $\alpha = 25°...75°$ and $\tan 25° \cong 0.46$; $\tan 75° \cong 3.7$ and remembering that $\tan(180° - \alpha) = \tan\alpha$ we have

$$\frac{l_y}{l_x} = 0.5 \ldots 3.5 \frac{[x]}{[y]}. \tag{7.24}$$

This should be observed, for adequate accuracy.

7.2.1.2 DEVICES USED IN GRAPHICAL METHODS

For the graphical evaluation of measurements, for constructions based on results arrived at by graphical methods (e.g., recordings), and for plotting tabulated information, the following simple devices may be used: compasses, calipers, rules, set squares (30 and 45°), drawing machines, curvilinear templates, scale rules, slide rules and simple electronic calculators. For graphical differentiation a mirror rule, a hard (3H) pencil and graph paper (simple mm, log, log-log, probability, frequency with probability, Gaussian, etc.) may be needed.

Further aids used in graphical methods – although less popular today – are some classical (mechanical) mathematical instruments. Their loss of popularity can be attributed to the emergence of the electronic computer. The electronic computer does not fully deserve this because in some instances these classical instruments serve the purpose in a simpler way, with the required accuracy, but at a lower cost. Space does not permit the author to go into the details, so he draws the reader's attention to References [7.1], [7.2], [7.3]. It is, however, worth noting that the instruments are built on a common principle: at one point in a mechanism fitted with arms and operating according to a suitable law of motion is a dial, called an integrating dial. The dial rotates at the plane of the paper and actuates a vernier counter. Since the integrating dial is attached to the mechanism, and is able to rotate, it causes the arm to perform a defined movement. If the movement of another point B in the same mechanism — in accordance with the law of the afore-mentioned motion, the dimensions, and the transmission characteristics of the integrating dial and the vernier counter – is appropriately coordinated with the angular displacement of the integrating dial then, while B goes around a plane figure (which may be irregular) the vernier counter gives a readout proportional to the area, the second degree moment, the static moment or some other quantity.

Several different mathematical instruments working on the same principle are available; the more important types will be dealt with below.

Planimeters

Planimeters have a single integrator dial. The two basic types of planimeter are polar and linear. A polar planimeter primarily determines areas enclosed by curves, i.e. computes integrals of the type $\oint y\,\mathrm{d}x$. A point C in its arm called the pole is fixed in the plane of the paper while point B moves around the plane figure being measured. A linear planimeter is used to determine areas under curves, i.e. to compute integrals of the type $\int y\,\mathrm{d}x$. A linear planimeter has no fixed poles: one end of the arm carrying the integrating dial runs on a straight track (hence the name) traces along the curve under study.

Other devices with a single integrating dial are the reciprocal and square-root planimeters, designed for the determination of integrals of the form $\int \mathrm{d}x/y$ and $\int \sqrt{y}\,\mathrm{d}x$.

A family of mathematical instruments operating on similar principles are the moment planimeters or integrators. They have 2, 3, or 4 integrating dials and are suitable for the determination of the static moment $\int y^2\,\mathrm{d}x$, the second-order moment $\int y^3\,\mathrm{d}x$ etc. $\left(\int y^4\,\mathrm{d}x, \int \sqrt{y}\,\mathrm{d}x, \int \sqrt{y^3}\,\mathrm{d}x\right)$. If the device not only has a vernier system to give a readout proportional to the integral but is also capable of plotting the curve which is proportional to the integral in question then it is called integraph (not very widespread as yet). Some other important mathematical instruments are the harmonic analysers. These can be used for the determination of the coefficients of graphically-given (true-to-scale) Fourier series of given periodic functions. Since such operations call for the computation of definite integrals, their use is also based on the movement of integrating dials but, for understandable reasons, their arm mechanism is more complicated than that of planimeters and integrators. In such devices it is of spatial character and includes not only integrating dials but also a glass integrating sphere.

To sum up: the (classical) mathematical instruments give valuable help in the graphical evaluation of measurement results. We have outlined their potentials. For their application, the reader should refer to their manuals and instructions for use.

7.2.2 BASIC TASKS IN GRAPHICAL MEASUREMENT EVALUATION

In the graphical evaluation of two- and multi-dimensional measurement information, the primary goal is the illustration to scale the interrelated measured data in the plane. If the information is two-dimensional, illustration in the plane is obvious: in the multi-dimensional case, the additional variables must be treated as parameters and for their illustration in the plane the use of the further variables as parameters is necessary. The points of the functional relation obtained by measurement, with certain errors, will represent the measurement information. Errors occur both when setting the independent variable(s) and when the function is being measured. If we can estimate the error limits for both the independent variable(s) and the function, then we can graphically show the upper and lower error limits.

Illustration of two-dimensional measurement information (function of a single variable)

The co-ordinate scales having been chosen, the measured co-ordinates (x_i, y_i) are plotted on the axes and then the points $P_i\ (x_i, y_i)$ so obtained are connected (Fig. 7.6). If we can estimate the absolute error limit involved when setting x and reading y, then we can graphically determine the two boundary curves which are the upper and lower error limits of $y(x)$.

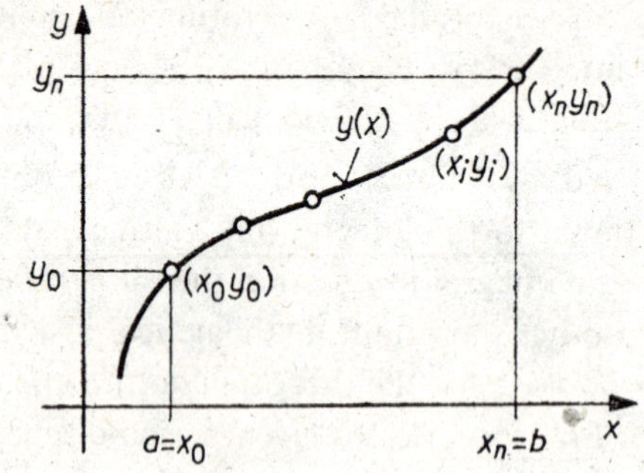

Fig. 7.6. *Functional curve with a single variable obtained by measurement*

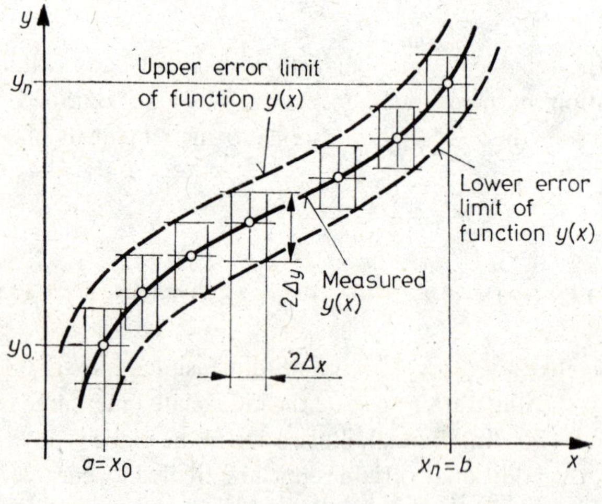

Fig. 7.7. *Geometrical determination of the upper and lower error limits of a function y(x) obtained by measurement*

Let Δ_x and Δ_y be the error limits of the co-ordinates in the measurement series. For the graphical determination of the two boundary curves, let us plot $y(x)$ once more (Fig. 7.7) and measure in both positive and negative directions the error limits Δ_x and Δ_y of the two co-ordinates, from every point of the curve. Then, connecting the corners of the small rectangular error areas in the figure, we get the upper and lower error limit curves of $y(x)$. The zone between the two limit curves is the tolerance zone, and may be regarded as the tolerance of the shape of the curve $y(x)$.

Illustrating multi-dimensional measurement information in the plane

In the multi-dimensional case the situation is more complicated. For instance, three-dimensional measurement information (a function of two independent variables) can be illustrated by a surface (Fig. 7.8). For illustration in the plane, parameters must be selected. It will thus be possible to illustrate in the plane the $z=f(x)|_{y=\text{const}}$

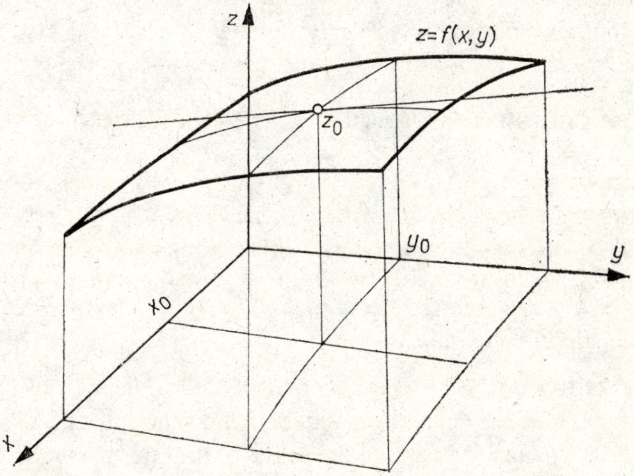

Fig. 7.8. *Geometrical illustration of three-dimensional measurement information*

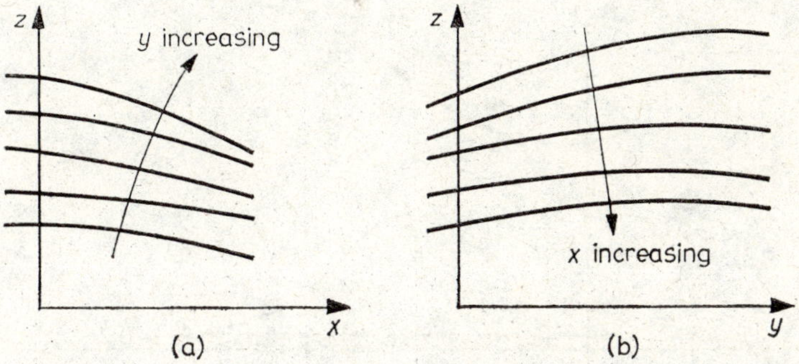

Fig. 7.9. *Illustration of the function according to Fig. 7.5 with the use of the further variables as parameters in the plane*

y=constant and $z=f(y)|_{x=\text{const}}$ x=constant sets of curves with a single variable. These are called the reference lines of the function which assume the forms seen in Figs 7.9a and b.

Four-, five-, etc. dimensional measuring information can be reduced in a similar way, step by step, to illustration in the plane. For the sets of curves which present themselves by the use of parameters, all the rules governing the two-dimensional information are applicable.

7.2.3 CONSTRUCTIONS, THE USE OF FUNCTION SCALES

In spite of the appearance on the scene of the electronic computer, the application of graphical methods is still advantageous in technical practice. We often need less accurate data, just for rough information, and there is neither need nor sufficient time for lengthy procedures. In such instances, graphical methods are invaluable. In the following section we shall briefly survey a number of graphical methods.

7.2.3.1 GRAPHICAL DIFFERENTIATION AND INTEGRATION

The analytical term for a curve or a set of curves plotted on the basis of measured points is "unknown". The data represented by the curve(s) is still "raw": in the course of evaluation it is subjected to a number of different mathematical (preferable graphical) operations, of which the most important are differentiation and integration.

Graphical differentiation

If the curve of the function $y=y(x)$ in the interval $a \leqq x \leqq b$ is given graphically in a co-ordinate system then the curve of its differential coefficient $y'(x)$ can be constructed, as in Fig. 7.10. Its curve consists of the values of the tangents to the function, plotted point-to-point (this refers to the method of plotting). A so-called pole point P is

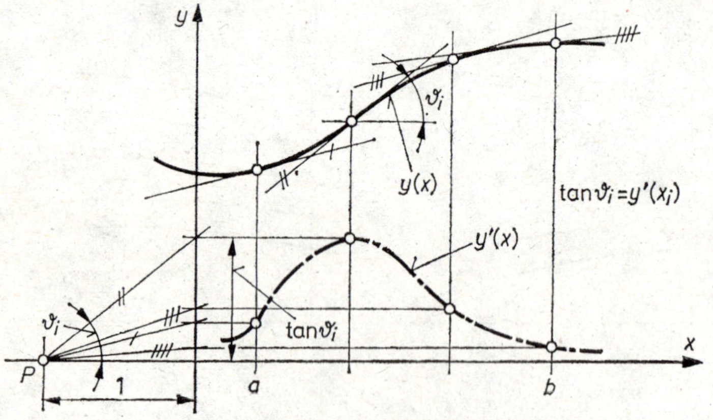

Fig. 7.10. *Graphical differentiation*

chosen at $(-1, 0)$ in the co-ordinate system, and lines are then drawn from P to the tangents to each point on the curve $y(x)$. These lines intersect the y-axis at the corresponding $\tan \vartheta$ where ϑ is the slope of the tangential line. Note that point $P\,(-1, 0)$ can seldom be on the same scale as the co-ordinate system (y, x), so it is preferable to work with an appropriate power of 10 (negative or positive) throughout.

Graphical integration

In many cases the integral curve* for the function $y=y(x)$ obtained empirically (by measurement) must be determined. If there is no analytical solution, a graphical method may be used (Fig. 7.11).

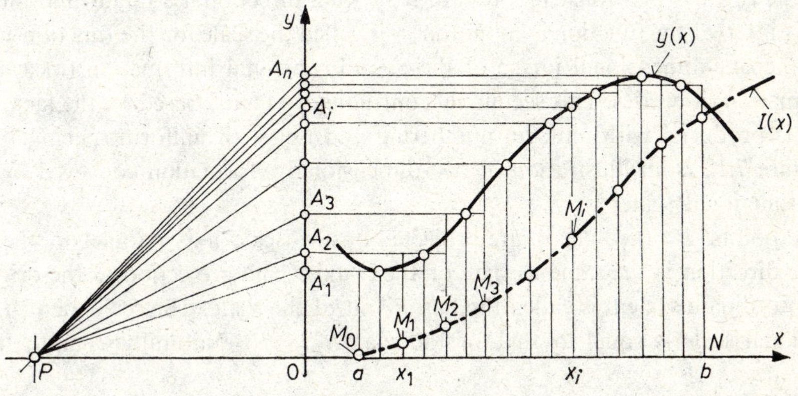

Fig. 7.11. *Graphical integration*

The interval (a, b) is divided into $2n$ (an even number) of equal parts by the points $a, x_{1/2}, x, x_{3/2}, x_2 \ldots x_{n-1}, x_{n-1/2}, b$ (a greater number of divisions ensures higher accuracy). At the midpoints the ordinate value of the integrand is projected onto the y-axis, to give the points $A_1, A_2 \ldots A_n$. Assume a pole point P on the negative part of the x-axis at a suitable distance, and connect it to points $A_1, A_2 \ldots A_n$. Now, with radii $\overline{PA_i}$ draw parallel lines from point a on the x-axis up to the edge of the interval (parallel with $\overline{PA_1}$ from x_1 to x_2), from the intersection point M_1 parallel with $\overline{PA_2}$ from x_1 to x_2. This will yield M_2, etc.) The areas of the sections of the polygon thus obtained are proportional to the area enclosed by the curve $y(x)$. It is important to know the factor of proportionality because on this depends the location of point P. If the ordinates of the polygon are denoted by $I(x)$ (to indicate that essentially an approximate integral curve is concerned), then we may write the following formula:

$$\int_a^{x_i} y(x)\,\mathrm{d}x \cong \overline{PO}\,I(x_i), \tag{7.25}$$

* Naturally not a primary function but a definite integral – an area below the curve – is concerned,

i.e., for the whole interval (a, b):

$$\int_a^b y(x)\,dx \cong \overline{PO}\,\overline{NM}_n, \qquad (7.25a)$$

which approximately yields the area below the curve of the graphically-defined function $y(x)$.

7.2.3.2 BASIC CONCEPTS OF FUNCTION SCALES

The functional values $y_i = y(x_i)$ corresponding to the abscissa values $x_i = x_0 + ih$ ($i = 0, 1, 2, \ldots, n$; h=const.) plotted to scale on a directional straight line interval— in this interval the function is monoton — is called the scale for the function $y = y(x)$. It is the one-dimensional image of the two-dimensional information (measurement information) it carries. This seemingly contradictory fact is based on the fact that the scale is produced with uniform numbering and not with uniform spacing.

Figure 7.12 is an illustration of two-dimensional information conveyed by means of a functional scale.

The points $P_i = ((x_i, y(x_i))$ are projected on the scale axis parallel to the x-axis whose direction is the same as that of the y-axis. The projection of the origin 0 of the co-ordinate system is called the base point of the scale. The scale factor l_s of the functional scale is equal to that of the y-axis, $l_s = l_y$. Naturally, l_s and l_y may be different.

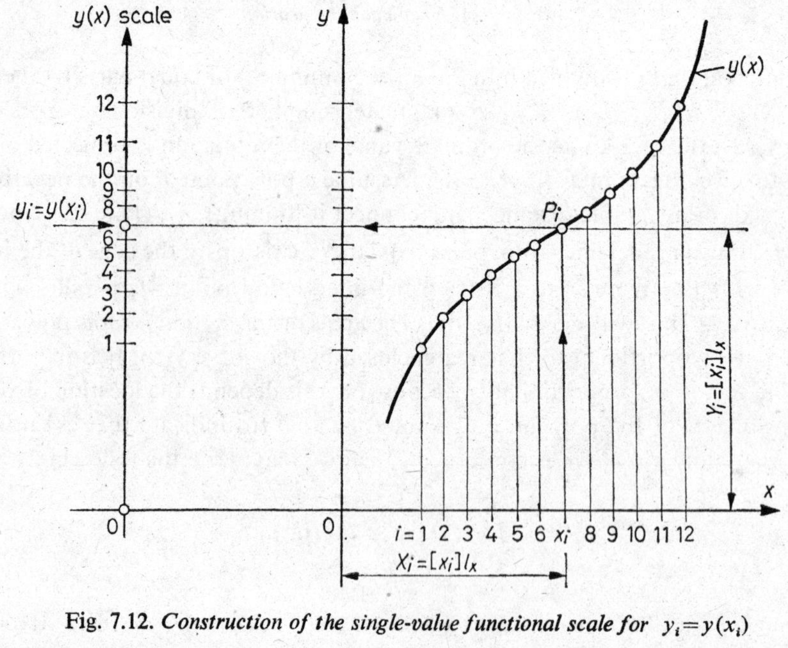

Fig. 7.12. *Construction of the single-value functional scale for* $y_i = y(x_i)$

The basic equation and scale ratio of functional scales

As shown by Fig. 7.12, the distances from the origin signed by the points are proportional to the corresponding functional values which they illustrate, and not to their numerical values. The correspondence of the points *P* with the numerical values to be shown on the scale in the interval depends on the values $y(x)$ of the function (Fig. 7.13).

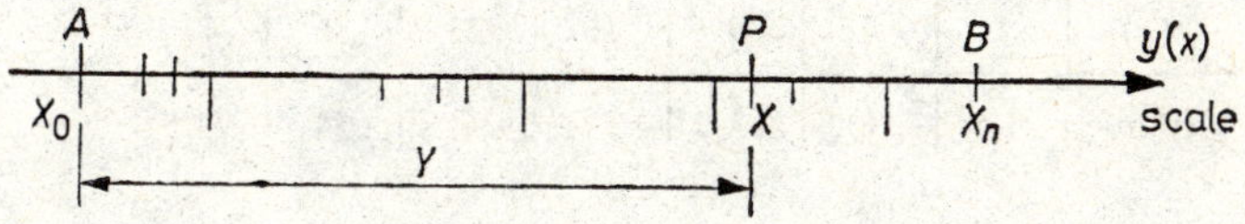

Fig. 7.13. *Illustration of the basic equation for function scales*

Making the point *A* the origin and determining the functional value $y(x_0)$ corresponding to it – but writing the value of x_0 at the point – and fixing the scale factor of the scale l_s, the distance $Y=\overline{AP}$ can be calculated from the relationship

$$Y = l_s[y(x)-y(x_0)]. \tag{7.26}$$

This is the basic equation of the functional scale. It helps to interpret the scale ratio l_s of the scale, i.e. the unit length of the steps in the function $y(x)$. Two points on the scale are at a distance l_s apart if the difference between the functional values corresponding to them is equal to 1. The scale factor can be determined if the two examined, extreme function values $y(x_0)$ and $y(x_n)$ are known. In Figure 7.13 the other point x_n has also been plotted. If the total length of the graduated scale must not exceed, say, the value *L* (the size of the sheet), by the appropriate use of (7.26) we may write that

$$l_s[y(x_n)]-y(x_0)] \leqq L \tag{7.27a}$$

which will make the required scale factor of the scale equal to

$$l_s \leqq \frac{L}{y(x_n)-y(x_0)}. \tag{7.27b}$$

In principle functional scales may be constructed for any function $y(x)$, but in practice, only the following are of major importance:

— power (root) function $\quad y(x) = x^n, y(x) = \sqrt[n]{x}$;

— logarithmic functions $\quad y(x) = \log x, y(x) = \ln x,$

$y(x) = {}^a\log x$;

— exponential functions $\quad y(x) = a^x, y(x) = e^x,$

$y(x) = 10^x$;

— trigonometric functions $\quad y(x) = \sin x, y(x) = \cos x,$

$y(x) = \tan x$, etc.

10*

Example

1. Plot the scale for the function $y(x)=x^2$ between $x=1$ and $x=8$. Let the length L that can be accommodated be 100 mm. In this case, the function for integral values of x is given in the following table:

x	1	2	3	4	5	6	7	8
$y(x)$	1	4	9	16	25	36	49	64
Y (mm)	1.5	6	13.5	24	37.5	54	73.5	96

$$l_s \leqq \frac{100}{64-1} = \frac{100}{63} = 1.5873\,\text{mm}.$$

With rounding downwards we might accept $l_s=1.5$. In the last row of the table we also show the distances Y, measured from the base point, obtained on the basis of the relation Y, mm$=l_s y(x)$.

Measuring all these distances and writing the numerical values of x beside the points, we arrive at the numerical form of the square scale (Fig. 7.14).

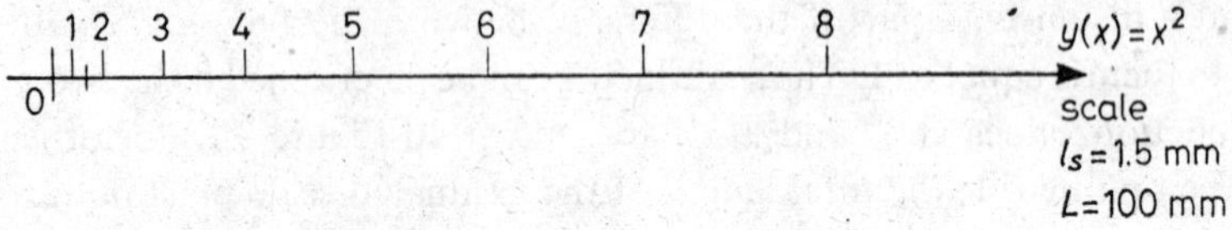

Fig. 7.14. *The scale* $y(x)=x^2$

2. Let us construct the scale for $y(x)=\cos x$ in the 0° to 90° range of the abscissa. As is well known in a right-angled triangle with a hypotenuse of unit length the cosine of an angle is equal to the length of the perpendicular side next to it. If, therefore, the graduations on an evenly-spaced (e.g. every 10) quarter circle of unit radius are projected onto a scale parallel to the perpendicular side next to the angle which is at the centre of the circle, and the graduations are written close to the points, we obtain the function scale for the function $\cos x$ (Fig. 7.15). So-called double scales may also be encountered, showing both x and $y(x)$. Such a double scale is shown in Fig. 7.16 for $\cos x$.

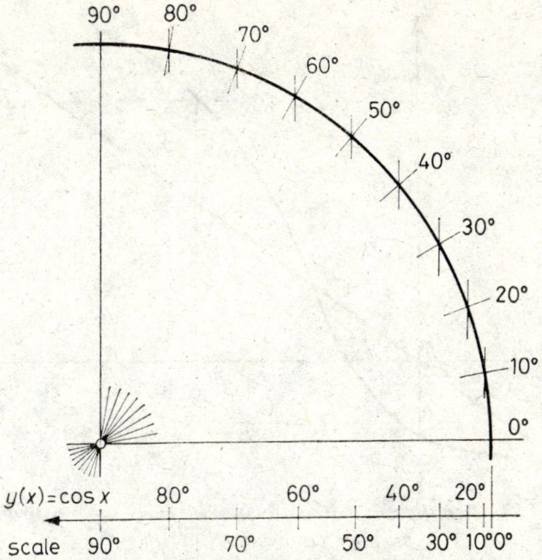

Fig. 7.15. *The scale for* $y(x)=\cos x$

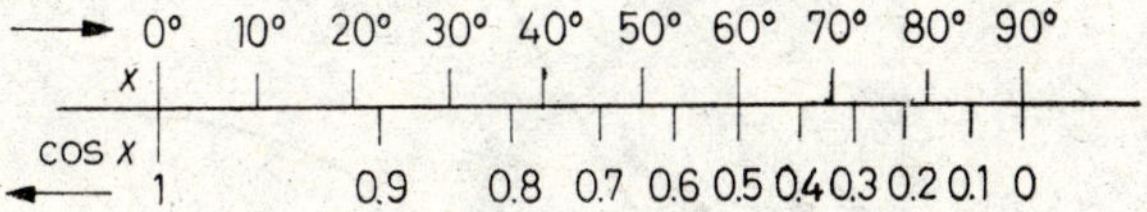

Fig. 7.16. *The double scale for* $\cos x$

7.2.3.3 CO-ORDINATE SYSTEMS WITH NON-LINEAR SCALES

Rectification

In Section 7.1.2 we have shown that by the introduction of new variables certain functions can be linearised. Such transformations can also be carried out graphically, if a function is given in graphical form by its measuring points. The curve of the graphically-illustrated function $y(x)$ in the co-ordinate system (x, y) converts into the straight line $\eta=m\xi$.*

(a) On a new axis ξ coincident with the x-axis we introduce a set of $\xi_i=\dfrac{1}{m}y(x_i)$ straight lines parallel with the ordinate defined by the divisions of the scale (with varying graduation) leaving the graduation of the y-axis unchanged: $\eta=y$ (Fig. 7.17) and

* In Section 7.1.2 the transformed functions are denoted by X and Y, as in literature. Here, however, the use of ξ and η is preferable because X and Y were, in graphical processes, used to denote other quantities: this could have caused confusion in the discussion.

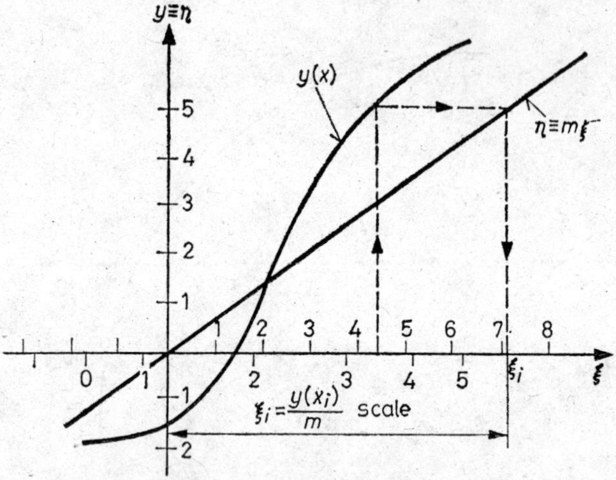

Fig. 7.17. *Rectification of a functional curve by an ordinate-parallel co-ordinate netting*

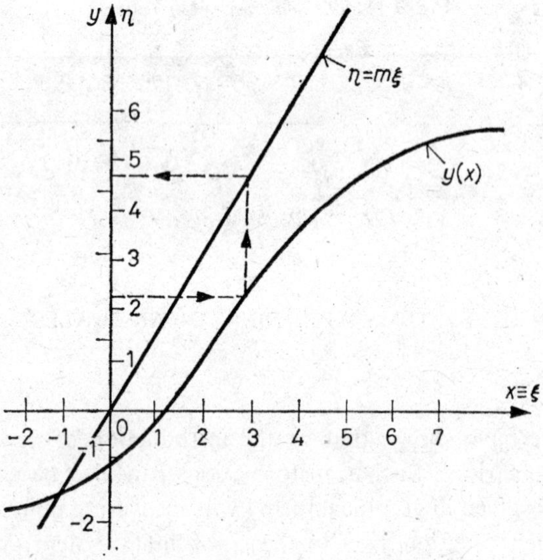

Fig. 7.18. *Rectification of a functional curve by an abscissa-parallel co-ordinate netting*

(b) on the axis coincident with the y-axis, a network of η_i straight lines parallel with the abscissa, defined by the divisions $\eta_i = m\dfrac{1}{x(y_i)}$ are introduced, leaving the spacing of the x-axis unchanged: $\xi = x$ (Fig. 7.18).

This explains how functional curves can be transformed into straight lines or into simpler curves by the use of square, logarithmic, etc. co-ordinate systems.

Graphical rectification

The basic task of equalization has been treated in Section 7.1.1.3. It is the construction of an approximate function whose difference from the measured values satisfies certain criteria. Using co-ordinate systems with varying co-ordinate spacing, the given function can be simplified (e.g. into a straight line). Plotting the measured points on this system will facilitate the drawing of the equalising straight line, even by eye.

Let the discrete measured values $y_i = y(x_i)$ be given. To simplify the function $y(x)$, the functions $\xi = \varphi(x)$ and $\eta = \psi(y)$ must exist, from whose substitution

$$\psi(y) = y[\varphi(x)] \tag{7.28}$$

we get the considerably simpler function

$$\eta = y(\xi). \tag{7.29}$$

For obvious reasons, the best case in practice is the one for which (7.29) yields a linear function. If two functions such as φ and ψ exist, then by scaling the co-ordinate axes accordingly, the curve will be straightened out.

Examples

In connection with the examples given below, we refer to Table 7.2. Using a suitable graduated co-ordinate system, we will demonstrate the method of rectifying a number of non-linear functions. In some cases we shall show only the technique, while in others we shall give a graphical illustration of the function.

1. In the graphical rectification of curves, logarithmic co-ordinates have considerable significance. We therefore show the exponential function $y = ab^{cx}$ first. With logarithms to base 10 of both sides we have

$$\log y = \log a + c \log bx.$$

Introducing the following notation:

$$\eta = \psi(y) = \log y,$$

$$\xi = \varphi(x) = k_2 x,$$

$$(k_1 = \log a; \quad k_2 = c \log b),$$

$$\eta = \psi(y) = y(\xi) = y[\varphi(x)] = k_1 + \xi = k_1 + k_2 x,$$

which shows that in an ξ, η co-ordinate system in which ξ is linearly and η logarithmically scaled, the original exponential function is shown by a straight line.

2. $y = ax^b$ is a power function. Again with the logarithm of the two sides

$$\log y = \log a + b \log x.$$

We now introduce

$$\eta = \psi(y) = \log y,$$

$$\xi = \varphi(x) = \log x,$$

$$(k_1 = \log a),$$

$$\eta = \psi(y) = y(\xi) = y[\varphi(x)] = k_1 + \xi = k_1 + b \log x,$$

which means that in a co-ordinate system with both axes logarithmic, a power function is shown by a straight line (on log–log paper).

3. Let us examine that part of an $\frac{x^2}{a^2}+\frac{y^2}{b^2}=1$ ellipse which falls into the first plane quadrant ($x \geqq 0$; $y \geqq 0$). Introducing the auxiliary functions $\eta=\psi(y)=y^2$ and $\xi=\varphi(x)=x^2$ the original ellipse turns into the straight line $\frac{\xi}{a^2}+\frac{\eta}{b^2}=1$

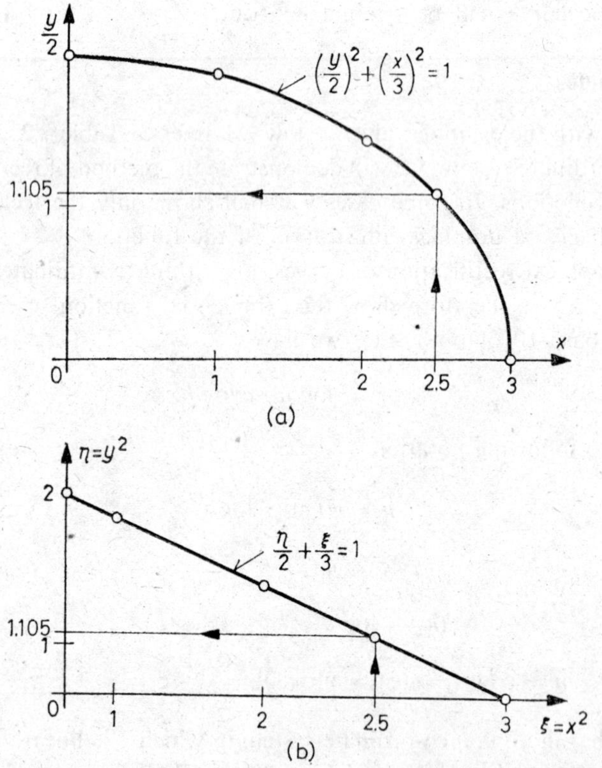

Fig. 7.19. *Rectification of the positive square of the ellipse* $\left(\frac{x}{3}\right)^2+\left(\frac{y}{2}\right)^2=1$ *by square graduated co-ordinate axes*

between the axes $+\xi$ and $+\eta$ (cf. the axial section, the so-called Salmon form of the equation of the straight line). Here the so-called axial sections are a^2 and b^2. If, therefore, instead of the linearly-graduated x- and y-axes we use the square-graduated ξ and η axes, the quarter-ellipse will appear as a straight line with negative slope. Let us show this with the following data: $a=3$ $(a^2=9)$, $b=2$ $(b^2=4)$. In Fig 7.19a we use a 3 cm unit scale and in (b) a 1 cm (this will clearly show that in this way the ellipse can be mapped into a parallelogram). If the points of the ellipse were obtained by a series of measurements, putting the points into the co-ordinate system $(\xi\ \eta)$, the straight (rectifying straight) line best fitting the points can even be plotted by eye.

4. The so-called polytropic relationship $yx^b=a$ (where $(pv^\varkappa=\text{constant})$ is of great importance in technical practice. With logarithms to base 10 of the two sides we have

$$\log y = \log a - b \log x,$$

whence we have the notation:

$$\eta = \psi(y) = \log y,\ \xi = \varphi(x) = \log x, \quad (k_1 = \log a)$$

which yields the straight line

$$\eta = k_1 - b\xi$$

showing that also the polytropic equation can also be rectified by the use of a log–log co-ordinate system (cf. case 2).

5. $y=Ke^{-(ax)^2}$ is the density function for the Gaussian probability distribution curve. This function, the so-called bell curve, can also be rectified. With the logarithms of the two sides we have

$$\log y = \log K - a^2 \log ex^2.$$

Further, with the notation $\eta=\log y$, $\xi=x^2$ $(k_1=\log k;\ k_2=a^2 \log e)$ we have

$$\eta = k_1 - k_2\xi,$$

which proves that the bell curve becomes straight in a co-ordinate system whose η-axis is logarithmic and whose ξ-axis is square. Let us take as a numerical example the equalization of the Gaussian curve where $K=2$ and $a=1$. Its equation is

$$y = \frac{2}{e^{-x^2}}.$$

In Figure 7.20a, the positive half of the bell curve is drawn with linear graduations, while in (b) the graduation is logarithmic-square. The straight line in the figure intersects the ξ-axis at approximately 0.75 — in good agreement with Fig. 7.19a, in which the bell curve also reaches unit value at around 0.75 and remains below 1. At $x=0.55$ its value is 1.55.

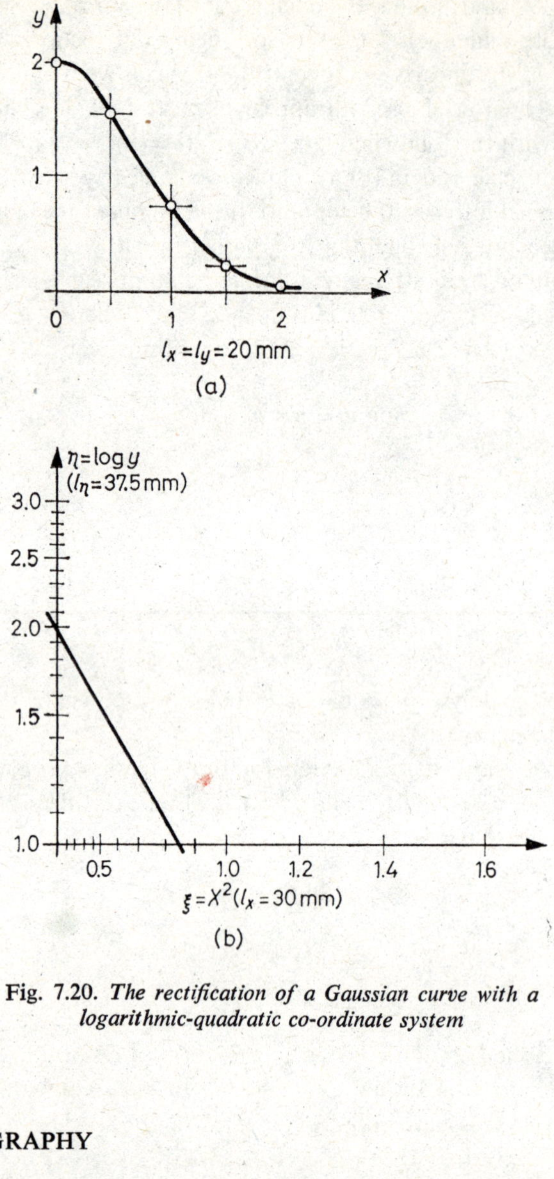

Fig. 7.20. *The rectification of a Gaussian curve with a logarithmic-quadratic co-ordinate system*

7.2.4 NOMOGRAPHY

We have proved in Section 7.2 that in processing one- and two-dimensional measurement information, graphical methods are nearly equivalent to the algebraic methods. Nomography (the use of alignment charts) is essentially the extension of graphical methods to multivariable functions, i.e. to multi-dimensional measurement information. In Section 7.2.2 it was shown that parametration allows the illustration of multivariable functions in the plane and from this aspect, the concepts of parametric illustration and nomography are akin. Nomograms are often called graphical tables.

What follows is a short overview of nomography and its application in the practice of measurement evaluation. Space does not permit the author to go into detail, so he calls the reader's attention to the relevant literature, and to Refs [7.2], [7.4], [7.6] and [7.7].

Basic concepts, the more important nomograms

Multivariable functions given analytically or by the co-ordinates of their measured points, can also be described by tables of alignment charts (nomograms). Nomograms present functions by point series, scales or curves marked with the numerical values of the variables, and by different methods of readout.

Two major groups of nomograms, and their combinations, are used:

— nomograms with sets of lines,
— nomograms with sets of points,
— their combinations.

Alignment charts (nomograms with sets of lines)

In such charts, each value of the function variables is represented by a line (straight or curved), and the different values of the variable are given by different lines or sets of lines.

Alignment charts can be constructed for all functions. Let us set out from the function $Z=f(x, y)$ and, by way of example, take the function $z=xy$ whose line chart is obtained by plotting the following sets of lines in the x, y co-ordinate system:

x=constant (straight line),
y=constant (straight line),
z=constant (hyperbola).

(Fig. 7.21), and write the z values not shown in the figure beside each curve. This will show that the curve $z=2\cdot5=10$ corresponds to the values $x=2$ and $y=5$; of the three sets of lines in this nomogram, two are straight and one hyperbolic, the last can be rectified by the simple means of taking the logarithms of $x=xy$ and getting

$$\log z = \log x+\log y \tag{7.30}$$

which proves that z=constant, and that the lines of constant $\log z$=constant are straight.

However, the nomogram with a set of hyperbolas does not satisfy the criteria of nomographic practice because not only is it difficult to construct hyperbolas but also interpolation with hyperbolas is difficult and inaccurate. Therefore the axes in nomograms are constructed with straight lines. The processes described before make this easy.

In the following, we shall use co-ordinate systems with axes graduated according to function scales, and see the way in which nomograms yield straight lines.

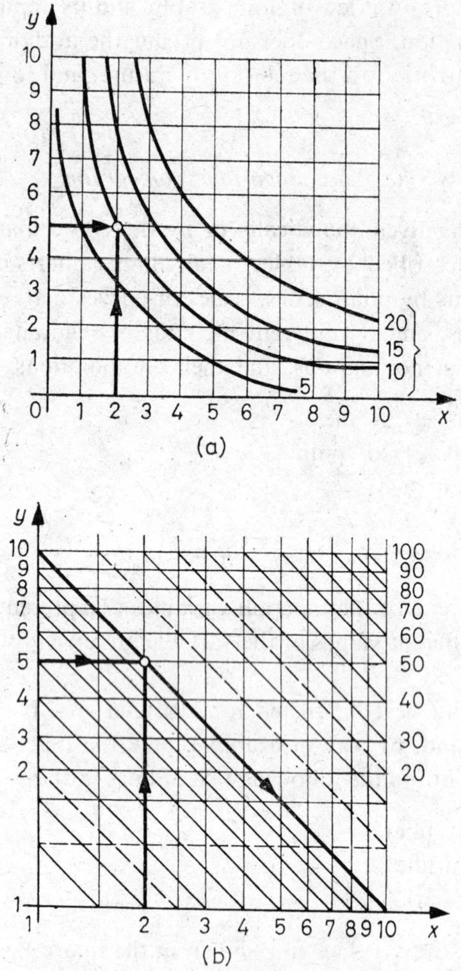

Fig. 7.21. *The nomograms of the set of lines* $z=xy$*:* (a) *hyperbolas,* (b) *rectified logarithmic variant*

Let a straight line be given in a co-ordinate system (Fig. 7.22) with arbitrary function scales for axes. The relation is given by the straight line between X and Y, up to point P, for which we define the relationship in the form of the well known equation of the straight line:

$$Y = MX+B \tag{7.31}$$

applying to this system.

Equation (7.31) having two parameters M and B, the parametric sets of straight lines may be obtained from the equations

$$Y = M_i X+B \tag{7.32a}$$

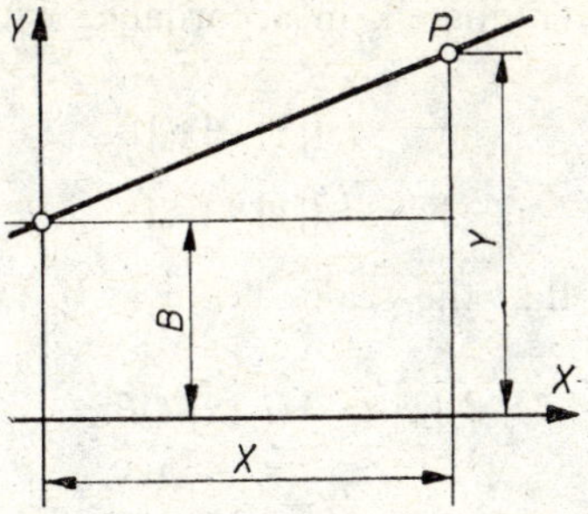

Fig. 7.22. *The interpretation of the basic concept of an alignment chart*

and

$$Y = MX + B_i; \qquad (7.32b)$$

(see Fig. 7.23).

The straight lines in equation (7.32*a*) and Fig. 7.23a are called a series of radii, while those in equation (7.32*b*) and Fig. 7.23b are called a series of parallel lines.

The number of function scales applicable to co-ordinate axes is limited. The most important ones, which help resolve a multitude of practical problems, are as follows:

— both axes are evenly graduated
— one axis is evenly and the other logarithmically graduated (log–normal co-ordinate system)
— both axes are logarithmically graduated (log–log co-ordinate system).

Now we shall try to ascertain the relationship between the physical quantities $x, y, z, \ldots$, their numerical values $[x]$, $[y]$, $[z]$ and their lengths $X, Y, Z, \ldots$ in the drawing.

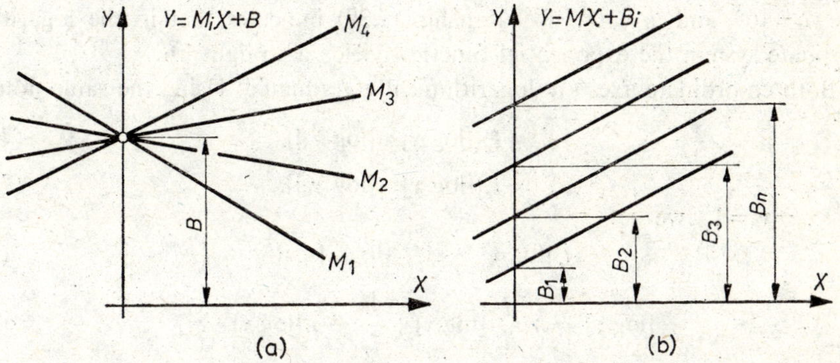

Fig. 7.23. *Two basic variants of an alignment chart:* (a) *set (series) of radii,* (b) *series of parallel lines*

1. Both axes are evenly graduated, in accordance with the earlier notations (cf. formula (7.26))

$$X = l_x([x]-[x_0]), \tag{7.33a}$$

$$Y = l_y([y]-[y_0]). \tag{7.33b}$$

We may generally assume that the scales begin at 0, i.e. $x_0=y_0=0$. Substituting (7.33*a* and *b*) into (7.31):

$$l_y[y] = Ml_x[x]+B. \tag{7.34}$$

Dividing through by l_y and introducing the notation $m=M\frac{l_x}{l_y}$ and $b=\frac{B}{l_y}$ we get

$$[y] = m[x]+b. \tag{7.35}$$

Equation (7.35) refers to the numerical values, (7.31) to the lengths in the drawing.

2. One axis is evenly and the other logarithmically graduated. When the abscissa is linear and the ordinate logarithmic then the equations of the scales are

$$X = l_x([x]-[x_0]) \tag{7.36a}$$

and

$$Y = l_y([\log y]-[\log y_0]). \tag{7.36b}$$

Substituting these equations with $[x_0]=0$ and $[y_0]=1$ into (7.31) we obtain

$$l_y[\log y] = Ml_x[x]+B, \tag{7.37}$$

and

$$[\log y] = M\frac{l_x}{l_y}[x]+\frac{B}{l_y} = c[x]+d, \tag{7.38}$$

having used the notation $c=M\frac{l_x}{l_y}$ and $d=\frac{B}{l_y}$. Making use of the well-known formulae of the logarithm, and by simple transformations, we can prove that

$$[y] = 10^{c[x]+d} = 10^d\cdot 10^{c[x]} = ba^{[x]}, \tag{7.39}$$

where $a=10^c$ and $b=10^d$. Relationship (7.39) indicates that in the logarithmic co-ordinate system the exponential function yields a straight line.

3. Both co-ordinate axes are logarithmically graduated. Using the same notation

$$X = l_x([\log x]-[\log x_0]), \tag{7.40a}$$

$$Y = l_y([\log y]-[\log y_0]). \tag{7.40b}$$

With $x_0=y_0=1$, we get

$$l_y[\log y] = Ml_x[\log x]+B, \tag{7.41}$$

$$[\log y] = M\frac{l_x}{l_y}[\log x]+\frac{B}{l_y} = n[\log x]+c, \tag{7.42}$$

where $n=M\frac{l_x}{l_y}$ and $c=\frac{B}{l_y}$.

Again using the formulae of the logarithm and performing simple transformations, we arrive at

$$[y] = 10^{n[\log x]+c} = (10^{[\log x]})\, 10^{c} \tag{7.43a}$$

which through the identity $10^{[\log x]+c} = x$ and the notation $10^c = b$ give:

$$[y] = b[x^n) \tag{7.43}$$

a power function, which shows that in the log–log co-ordinate system the power function is represented by a straight line.

We now revert to the alignment charts. Obviously, if e.g. in the two-variant function concerned z is a multiplier the result will be a set of radii, while if z exerts an influence on the intersect the result will be a set of parallel lines.

Let us now examine what substitutions are required to obtain nomograms with sets of straight lines, sets of radii or sets of parallel lines in the three co-ordinate systems, i.e. normal, log–normal and log–log.

1. If both axes are evenly graduated, in (7.35) the following substitutions are necessary:

To obtain a nomogram of parallel lines, in order to illustrate (7.35) we have to substitute $m = c_1$ and $b = c_2[z]$ to get the relationship

$$[y] = c_1[x] + c_2[z], \tag{7.44}$$

which is a series of straight lines. For obtaining a series of radii, then substituting $m = c_1[z]$ and $b = c_2[z]$ the variable z is considered in the slope:

$$[y] = c_1[x][z] + c_2. \tag{7.45}$$

2. One axis (the abscissa) is evenly, the other (the ordinate) logarithmically graduated. The following substitutions into (7.39) are required: If parallel lines are to be obtained, then substituting $b = c[z]$ we get

$$[y] = ca^{[x]}[z], \tag{7.46}$$

while if we want to have a set of radii, then substituting $a = [z]$ and $b = c$, we have

$$[y] = c[z]^{[x]} \tag{7.47}$$

both of which satisfy the requirement.

3. Both axes are logarithmically graduated. Now we have to substitute into Eq. (7.43). To get parallel lines we substitute $b = c[z]$ and get

$$[y] = c[x^n][z] \tag{7.48}$$

for a set of radii, substituting $b = c$ and $n = [z]$ we have

$$[y] = c[x^n], \tag{7.49}$$

which are really parallel straight lines or series of radii.

Summing up, alignment charts can be constructed for all kinds of functions with several variables. Among these, the most important in practice are those nomograms which yield sets of straight lines, because straight lines are easier to plot and to handle.

Nomograms with series of points

These nomograms combine several function scales ($x, y, z, \ldots$). Their main feature is that the value of the variable is determined by a point, as opposed to a line. The nomogram which consists of scales, may be straight or curved lines but for easier handling, preference is given to those which yield straight lines. In the nomography of function scales are often called holders. In the nomographic illustration of function relationships with three variables $z(x, y)$, Fig. 7.24 shows the most frequently used forms – with straight function scales:

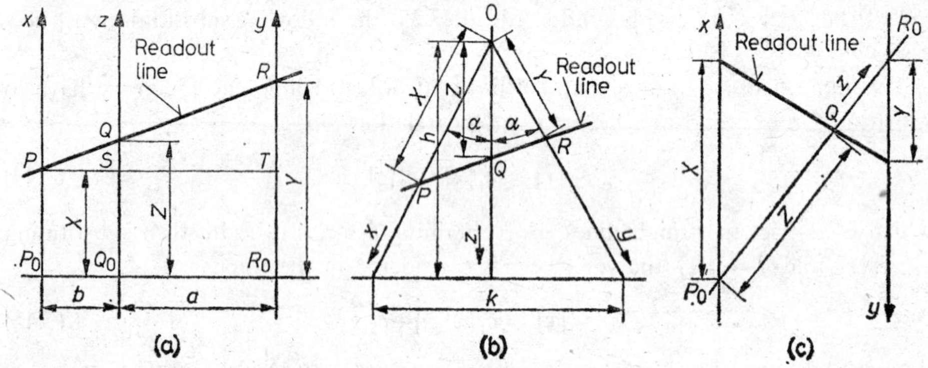

Fig. 7.24. *The nomograms most frequently used to illustrate function relationships of the* $z(x, y)$ *type* (a) *nomogram with parallel function scales,* (b) *nomogram with function scales starting out from the same point,* (c) *nomogram with two parallel function scales intersected by a third (the so-called N-nomogram)*

By a suitable choice for the graduation of the function scales and their location, different function relationships can be satisfied. With the earlier notations, let us survey the more important basic relationships of the nomograms in Fig. 7.24.

On the basis of the similar triangles in Fig. 7.24a, the following proportionality applies to the nomogram with parallel function scales:

$$\overline{SQ}:\overline{TR} = \overline{PS}:\overline{PT}$$

or substituting the distances indicated in the figure:

$$(Z-X):(Y-X) = b:(a+b). \tag{7.50a}$$

After rearrangement and simplification, we have the basic equation:

$$Z(a+b) = aX+bY. \tag{7.50b}$$

If the functional scales have different functional divisions (e.g. logarithmic, reciprocal, square root, trigonometric, hyperbolic etc.), i.e. if they are scaled according to $f_1(x)$ instead of x, $f_2(y)$ instead of y and $f_3(z)$ instead of z then, from (7.26), the basic scale equations will be

$$X = l_x[f_1(x)-f_1(x_0)], \tag{7.51a}$$

$$Y = l_y[f_2(y)]-f_2(y_0)], \tag{7.51b}$$

$$Z = l_z[f_3(z)-f_3(z_0)]. \tag{7.51c}$$

Substituting these into (7.50b) we get the so-called key equation of the nomogram with parallel scales.

Without detailed calculations, the key equation is as follows

$$Cf_3(z) = Af_1(x)+Bf_2(y). \tag{7.52}$$

We introduce the simplifying notation:

$$A = al_x, \ B = bl_y \ \text{ and } \ C = (a+b)\,l_z \tag{7.53a—c}$$

and made the origins of the scales equal to their zero points, i.e.

$$f_1(x_0) = f_2(y_0) = f_3(z_0) = 0.$$

For simplicity, the nomogram with function scales starting from a single point, as in Fig. 7.24b, was chosen so as to be symmetrical. The geometric relationship from which we start in this case is based on the sum of the areas of the two triangles above the readout line. It may be written

$$T_{\Delta \mathrm{OPR}} = T_{\Delta \mathrm{OPQ}} + T_{\Delta \mathrm{OQR}},$$

i.e.

$$\frac{1}{2}XY\sin 2\alpha = \frac{1}{2}XZ\sin\alpha + \frac{1}{2}YZ\sin\alpha. \tag{7.54}$$

Dividing Eq. (7.34) by $\dfrac{1}{2}XYZ\sin\alpha$:

$$\frac{\sin 2\alpha}{Z\sin\alpha} = \frac{1}{X} + \frac{1}{Y}, \tag{7.55a}$$

or, making use of the relationship $\sin 2\alpha = 2\sin\alpha\cos\alpha$

$$\frac{2\cos\alpha}{Z} = \frac{1}{X} + \frac{1}{Z}. \tag{7.55b}$$

The scale of the two outer function scales will be identical, i.e. $l_x = l_y = l$, and assuming that the middle one is graduated according to $l_z = 2l\cos\alpha$ and furthermore that the functional scales start at zero, then, substituting

$$X = lf_1(x), \tag{7.56a}$$

$$Y = lf_2(y), \tag{7.56b}$$

$$Z = 2l \cos \alpha f_3(z), \tag{7.56c}$$

the scale equations into (7.35), the key equation will become

$$\frac{1}{f_3(z)} = \frac{1}{f_1(x)} + \frac{1}{f_2(y)}. \tag{7.57}$$

The function scale x of the N-nomogram in Fig. 7.24c starts at the point P_0 and its function scale at the point R_0, while the function scale z starts at the point P_0 and ends at R_0 with length $P_0R_0 = a$. The readout line cuts off sections of X, Y and Z. The basic relationship is obtained from the proportionality of the two similar triangles:

$$X:Y = Z:(a-Z) \tag{7.58}$$

Let the arbitrary graduations on the functional scales x and y be

$$X = l_x f_1(x), \tag{7.59a}$$

$$Y = l_y f_2(y). \tag{7.59b}$$

If the function scale also had an arbitrary graduation, the result would be the key equation for a special case. It is therefore preferable to replace the right side of (7.58) by

$$\frac{Z}{a-Z} = l_z f_3(z). \tag{7.60}$$

Solving for Z:

$$Z = a\frac{l_z f_3(z)}{l_z f_3(z)+1} = a\frac{f_3(z)}{f_3(z)+\frac{1}{l_z}}. \tag{7.61}$$

If l_z is substituted by l_x/l_y then the key equation becomes

$$f_3(z) = \frac{f_1(x)}{f_2(y)}. \tag{7.62}$$

This would illustrate the quotient of the two functions. The problem could also be solved with three parallel function scales, but in that case it would be necessary to use the logarithms of the equations, which could involve some problems (e.g. in compound nomograms).

Summing up: the nomograms with series of points provide an easy-to-handle means of illustration for many (but not all) multivariable functions. They help to evaluate and plot multi-dimensional measurement information. Before using them, it is advisable first to learn (with one of the methods previously outlined) about the character of the measurement information (e.g. tabulated multivariable function) and, on that basis, choose the type of the monogram best suited to the case in hand.

Compound nomograms

So far, functions with three variables (two independent) have been dealt with. If there is a function with e.g. four variables

$$v = f(x, y, z). \tag{7.63}$$

then, by introducing a suitable auxiliary variable u, part of the function:

$$u = f_1(x, y) \tag{7.64}$$

can be (usually but not always) removed from the original function (7.63). In this way it will change to

$$v = f_2(z, u). \tag{7.65}$$

Including the auxiliary variable, we thus have five variables in the two equations. The two part nomograms can now be handled by one of the previously described methods – regardless of whether the nomogram is an alignment chart or one of point character.

Here are some basic rules for the joining of two part nomograms.

1. If two function scales are indicated in direction, sense, division and scale, then they can be superimposed or placed side by side and a set of parallel lines take over the lead from the first to the second nomogram (Fig. 7.25), again irrespective of whether series of points or line sets are concerned.

2. If two function scales have the same direction, sense and the type of scale divisions but different scale factor, then central projection allows their linking together (Fig. 7.26). If the centre of the projection lies between the two function scales, even their direction may be changed.

3. If the divisions on the function scales are identical but their directions differ, they can be linked by parallel projection or by reflection at a so-called guide line AC (Figs 7.27a and b).

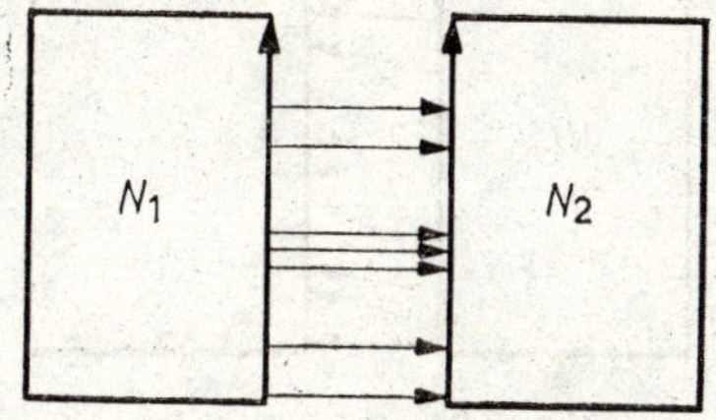

Fig. 7.25. *Two scales superimposed*

4. If the divisions on two function scales differ, they must be connected point-to-point (Fig. 7.28).

Finally, it should be noted that similar techniques are possible for linking alignment charts. Because of lack of space we do not go into details here but draw the reader's attention to References [7.2], [7.4], [7.6] and [7.7].

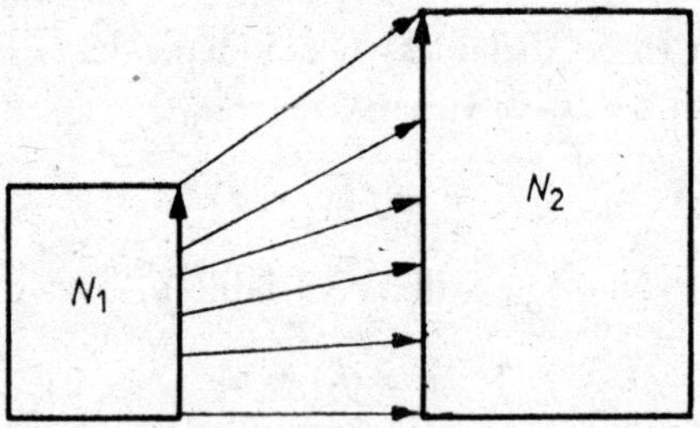

Fig. 7.26. *Two scales linked by central projection*

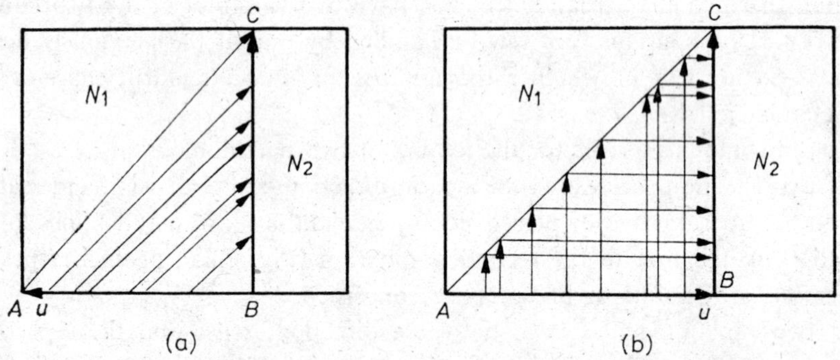

Fig. 7.27. *Linking nomograms if the divisions of the functional scales are identical but their directions are different:* (a) *parallel projection,* (b) *reflection at the guide lines*

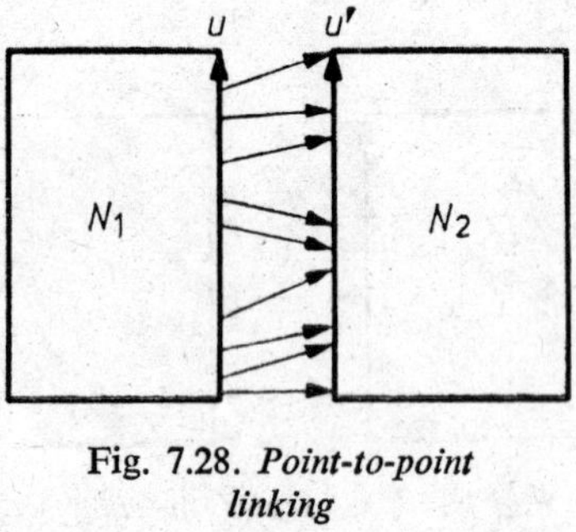

Fig. 7.28. *Point-to-point linking*

8. EVALUATION OF DETERMINISTIC MEASUREMENT INFORMATION. PART 3.

Evaluation by frequency; frequency analysis; spectral evaluation; methods of analysis

In the preceding chapters we dealt with the evaluation of measurement information and of signals, by value. Now we shall examine the evaluation of information carried by frequency.

8.1 BASIC CONCEPTS

Measurement information may often be latent, hidden in the frequency of the measured signal or in some other characteristic related to its frequency. The term "hidden" is justified, because the properties of the frequency that is, the information it carries, its handling, evaluation, etc. appear with less clarity than in the evaluation of information by value. Measurements yield such results from oscillations, the examination of automated systems, acoustical measurements, measurement of a.c. current and a.c. voltage, the analysis of time functions, etc.

Any arbitrary periodic function can be expressed by Fourier analysis as the sum of a finite number of discrete harmonic components or of an infinite number of non-periodic functions of different frequencies. If, in shock analysis, we examine a power impulse taking place according to a certain time function, its frequency composition will be of primary importance. Here is a more simple example. If the input of a control circuit is excited by a definite (harmonic, periodic or aperiodic) test signal, the response at the output depends on the properties of the instrument or the system being studied. This dependence manifests itself in the fact that the instrument responds differently to test signals of different frequencies or frequency compositions. This behaviour is called frequency dependence.

In connection with measurement information and evaluation of measurement information carried by frequency or frequency properties, there are several important methods and processes:

— for the examination of the frequency dependence of instruments:
  (a) examination with harmonic test signals,
  (b) examination with other test signals.
— for the analysis of signal frequencies
  (a) octave- and third-analysis,
  (b) narrow-band analysis.

Although in practice diverse physical quantities are examined (displacement, power, pressure, sound pressure, flow, etc.) all of which would require separate treatment, we shall concentrate on the frequency analysis of electrical signals alone. This is because the physical quantities on which frequency analysis is performed are, almost without exception, measured by electrical means (by resistive, capacitative, inductive, piezoelectrical, thermionic, Hall-generator etc. devices). Furthermore, as the reader will see, examinations by frequency cannot be performed in any other way, because of the high speed required, the complexity of the process, and some other reason. We may add the important (empirical) fact that the physical quantities to be submitted to frequency analysis occur at natural frequencies which make electrical instruments indispensable. In the circumstances, the commercially available and generally used instruments of frequency analysis use electrical signals almost exclusively. So we, too, shall concentrate on such means and from time to time refer also to the non-electrical/electrical interfaces which link them.*

Measurement information carried by an electric signal can be evaluated on the basis of the signal's frequency properties:

— by a measured signal taken off directly the measurand object (mostly non-repeatable),
— by intermediate analogue storage (mostly on magnetic tape).

The two processes will be dealt with in more detail later on in this chapter. In what follows we shall survey the processes and means used in the examination and evaluation of frequency properties.

8.2 EXAMINATION AND EVALUATION OF THE FREQUENCY-DEPENDENT BEHAVIOUR OF MEASURING INSTRUMENTS

The examination of the frequency-dependent behaviour of the elements, smaller or larger parts etc., of instruments or equipment is an everyday problem not only in automation but also in many other fields – including measurement technique. As instrument, system and subsystem may be electrical, electronic, mechanical, pneumatic, hydraulic, acoustic, thermal, etc., or a combination of these. It could be a simple member of an electrical RC (Resistive–Capacitive), a mechanical oscillating circuit, an electronic amplifier, a heat-storage unit, a hydraulic amplifier, an acoustic transmitter, etc. A detailed examination of these alternative possibilities would go beyond the scope of this book, but there is copious literature on the subject. We shall

* The English term "interface" has been accepted (mainly in computer applications) for mechanical-, electrical-, pneumatic-, hydraulic-, electrical-pneumatic, etc. devices which link parts of systems. Although in the Hungarian language the term "transducer" would be more appropriate, in consideration of the general acceptance of the English term, the linguistic difference has been ignored.

give here a brief survey of the measurement methods which are common in the examination of the frequency-dependent behaviour of such instruments.

Knowledge of its frequency-dependent behaviour is necessary to predict the dynamic response of the device concerned, and to design the complete system in which it is to operate. Obviously, to predict the likely response of a system, the behaviour of its elements and subsystems must be known.

During normal operation, the input of the device is affected by the so-called input signal. The device responds to this by generating another effect, at its output, the so-called output signal.

The output signal – the response – depends on the input signal and on the characteristics of the device. Since we are interested only in the effect of the properties of the device, we disregard the properties of the input signal and confine our examination to a few easy-to-handle signals, known as typical test signals. The most frequently used typical test signals have been compiled in Table 8.1.

As is known, the connection between the input and the output of the device may be linear or non-linear. These terms refer to their mathematical description. Let us assume that the devices to be examined (and thus the systems into which they are built) are linear.

Another property of the connection between the input and output of the device is the frequency dependence. Frequency-dependent transmission means that the device responds differently to input signals of different frequencies.

For example, when a signal with several frequency components is transmitted, the device will transmit some components very well, others weakly and some it will suppress altogether.

The test signals in rows 1–4 of Table 8.1 are used in the examination of the transient behaviour of elements, devices, or systems. For the modelling or the examination of steady state conditions, the harmonic functions in the 5th column should be used.

Devices may be passive or active.*

Figure 8.1 illustrates the general arrangement of the examination of frequency dependent behaviour. For this purpose a unit is required which can produce the necessary continuously-adjustable test signals in the desired frequency range, together with a device to measure the output signal connected by feed forward to the unit which produces the test signal. This is necessary to compare the input and output signals, since for evaluation their amplitude and phase relations must be known.

Accordingly, feeding the test signals to the input of the device, the signal at the output must be measured. In the case of examination of transient this would require the recording of the output signal generated by a single test signal, while in the examination of transmission in the steady state, it would mean a series of measurements within the frequency range concerned, depending on the frequency.

* A device is passive or active depending on whether it contains an energy source. A passive device, consuming energy, attenuates the input signal, while an active device amplifies it.

TABLE 8.1.

Typical test signals

Type of test signal and its time function	Form of the time function	Remarks
1. Unit pulse function (*Dirac*-delta) $\delta(t)$	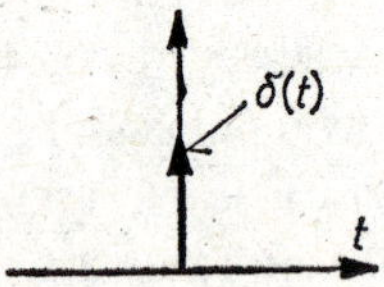	Interpretation: if $t=0$, $\delta(t)=\infty$; if $t \lessgtr 0$; $\delta(t)=0$ In a linear system the response to $\delta(t)$ is the weight function $\int_{-\infty}^{\infty} \delta(t) = 1$ 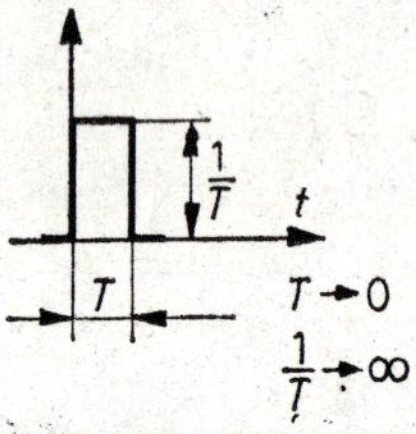
2. Unit step $1(t)$	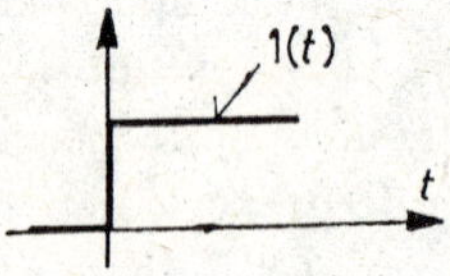	Interpretation: $t < 0$; $1(t) = 0$ $t \geqq 0$; $1(t) = 1$ In a linear system the response to $1(t)$ is the transfer function
3. Function of unit velocity step $t1(t)$	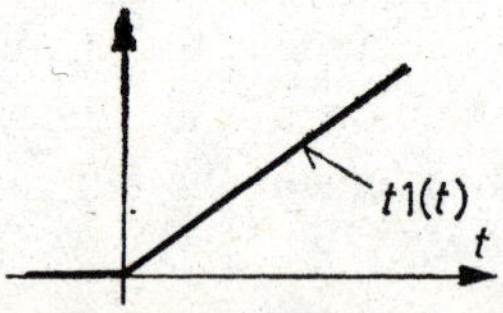	Interpretation: $t < 0$; $t1(t) = 0$ $t \geqq 0$; $t1(t) = t$ The response is the transfer function referring to the unit velocity step
4. Function of unit acceleration step $t^2 1(t)$	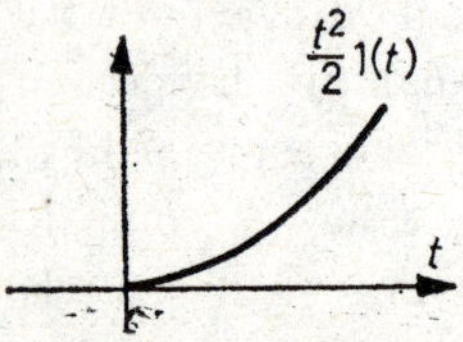	Interpretation: $t < 0$; $(t^2/2)1(t) = 0$ $t \geqq 0$; $(t^2/2)1(t) = t^2/2$ Response: the transfer function referring to the unit acceleration step
5. Harmonic function $\sin(\omega t)$ $\cos(\omega t)$	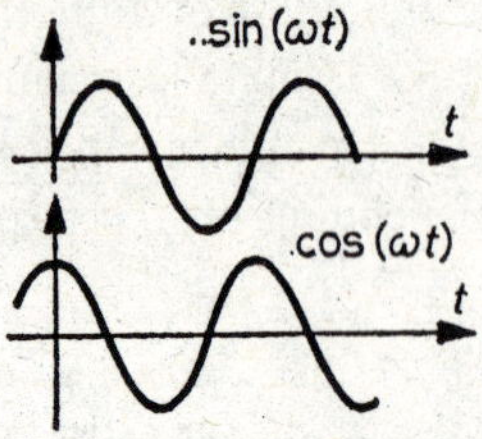	

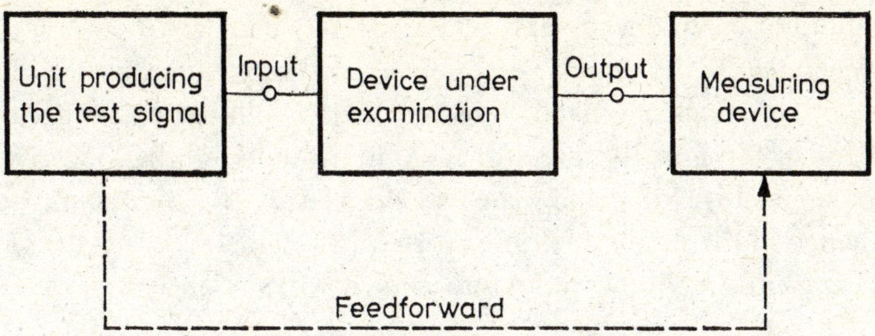

Fig. 8.1. *Examination of frequency dependence; amplitude vs. frequency*

8.2.1 EXAMINATION WITH A HARMONIC TEST SIGNAL

Feeding a sine (or cosine) test signal into the input of the device, we obtain different output signals at different frequencies, depending on its frequency-dependent properties, i.e. at different frequency values the device behaves differently. This, in a nutshell, is frequency dependence, and our examination consists of the comparison of the amplitude and phase of the input and output signals at each frequency. Let the input signal to the device be $x_1(t)$ and its output signal $x_2(t)$. The relation between the two signals characterises the signal transfer of the device. Assume further that the device is linear and that its input is excited by a harmonic time function. In computations, harmonic time functions are replaced by complex rotating vectors on conditions that in the course of further computations their real or imaginary components should be taken into consideration.

The input signal accordingly is

$$x_1(t) = |A|e^{j(\omega t+\varphi_1)}. \tag{8.1}$$

If the device is linear then it has no effect on the angular frequency ω of the harmonic signal, but it causes the amplitude to change (damps or intensifies it), and its phase to shift. Thus, at the output of the device the harmonic signal

$$x_2(t) = |A|e^{j(\omega t+\varphi_2)} \tag{8.2}$$

will be measured. The signal transfer of the device is characterised by the quotient

$$Y = \frac{x_2(t)}{x_1(t)} = \frac{|B|}{|A|}e^{j(\varphi_1-\varphi_2)} = Ke^{j\varphi}. \tag{8.3}$$

Naturally $K=\dfrac{|B|}{|A|}$ and $\varphi=\varphi_2-\varphi_1$.

Both K and φ being frequency dependent, $K=K(\omega)$ and $\varphi=\varphi(\omega)$. To ascertain the frequency-dependent properties of the device, the functions $K(\omega)$ and $\varphi(\omega)$ must be determined.

In the amplitude vs. frequency and phase vs. frequency charts which characterise frequency dependence, the angular frequency ω is substituted by $\omega=2\pi f$ and to illustrate the broad frequency band, f in the chart is plotted on a logarithmic scale. It is also customary to give $K(\omega)$ and/or $K(f)$ in dB units, so that negative dB values mean damping, positive dB values mean gain and 0 dB values stand for transfer without a change in the amplitude (undistorted). In Table 8.2 some dB values corresponding to certain amplitude ratios are shown for guidance.

TABLE 8.2.

The numerical values of some amplitude ratios in terms of dB

Numerical value of amplitude ratio K	Amplitude ratio K (dB)
0.1	−20
0.2	−14
1	0
2	+6
10	+20
100	+40

The amplitude vs. frequency and phase vs. frequency curves for a device, recorded on a logarithmic scale are also called Bode diagrams.

Bode, an American researcher first used them to describe the frequency-dependent properties of devices (see Ref. [8.1]).

8.2.2 EXAMINATION WITH OTHER TEST SIGNALS

In the preceding section we showed the method of determining the frequency-dependent properties of devices using harmonic test signals. The information gained is characteristic primarily of operations under steady state conditions. Other test signals shown in Table 8.1 may also be used for examination, in the manner shown in Fig. 8.1. In such cases the unit producing the test signal generates the desired signal (e.g. a step function). The output signal so obtained will be characteristic of the transient operation of the device. The output signal produced by the test signal is recorded, because its trend provides important information about the transient operation of the device. We cannot go into more detail on transient analysis, but wish to note that from among the test signals in Table 8.1 the Dirac delta is used mostly in theoretical examinations, while the unit step can be used to advantage in practice, and in experiments, because it is easy to produce by making a switch.

The functions unit velocity jump and unit acceleration jump can be produced by special function generators, for unusual examinations.

8.3 DEVICES USED IN FREQUENCY ANALYSIS

As referred to in Section 8.1, certain pieces of measurement information are carried in the spectral properties, frequency composition or other frequency-dependent parameters of the signal. When the information is carried by the spectral properties of the signal then the carrying signal must be subjected to frequency analysis (spectral examination). Frequency analysis will provide information not only about the greatest amplitude and the amplitude level of the signal but also about the amplitude levels of the components of the different frequencies, separately.

Steady state signals can be analysed by frequency analysers (known also as selective amplifiers). They consist of an electronic amplifier and tunable filters which let only a narrow frequency band pass and suppress the frequency components above and below it.

Transient signals (as in e.g. shock analysis) can be analysed by a real time analyser.

In what follows, first the devices used in frequency analysis will be discussed, and then its methods. Since filters are the most important devices in frequency analysis, analysers are built up from filters.

8.3.1 FILTERS

Filters are circuits including passive – more lately also active – electrical (electronic) components with pass band and non-pass band frequency ranges. The attenuation of filters is high in the non-pass band and low in the pass band, so they effectively let pass only those signal components whose frequency falls within the pass band. "Attenuation" characterises the "power suppression" of the filter (more precise definitions follow later), [8.3], [8.4].

In frequency analysis carried out directly from the measured signal with intermediate analogue memory the filter must be inserted between the memory and the indicator.

Here follows a survey of some important problems in the theory and the application of filters.

Matching and attenuation

Filters are quadrupoles which match the power between the signal source and the load. In frequency analysis, the signal source emits the signal to be examined if intermediate analogue storage takes place between the signal-reproducing storage and the indicator (pointer) or recorder. The indicator generally applies an ohmic load to the filter. The arrangement of the filter is shown in Fig. 8.2.

If, as desired, the filter is loaded and matched on both sides, i.e. at its input and output, then the loaded resistances are equal to the input and output resistances, i.e. $R_{in}=R_{inlet}$ and $R_l=R_{outlet}$, and maximum power can be obtained from the driving

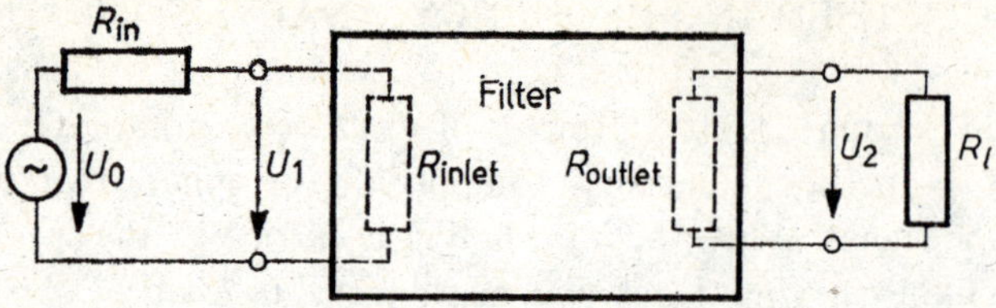

Fig. 8.2. *Arrangement and operation of a filter circuit: U_0 — the source voltage of the signal source, R_{in} — the internal resistance of the signal source, R_1 — the load (internal resistance of the indicator), R_{inlet} and R_{outlet} — the input and output resistances of the filter, U_1 and U_2 — the input and output voltages of the filter*

generator. Its notation is

$$P_0 = \frac{|U_0|^2}{4R_{in}}. \tag{8.4}$$

In this case the effective power across the load R_l which loads the filter output is

$$P_2 = \frac{|U_2|^2}{R_l}. \tag{8.5}$$

To characterise the transfer of the filter, a so-called transfer factor Γ is defined. It shows the transfer and damping properties of the filter. The transfer factor of the filter, more precisely its square, is defined by the quotient of the power at the input side and the actual power at the output side given by (8.4) and (8.5) respectively

$$|\Gamma|^2 = \frac{P_0}{P_2} \geqq 1. \tag{8.6}$$

This stipulation facilitates computation with logarithmic (Np and dB) units.

It should be noted that in telecommunications neper or decibel units are usually used, while in measuring techniques the decibel is used almost exclusively.

The transfer and attenuation properties can be derived from (8.6). To do so, let us first substitute (8.4) and (8.5) into (8.6):

$$|\Gamma|^2 = \frac{P_0}{P_2} = \frac{|U_0|^2 R_l}{|2U_2|^2 R_{in}}; \tag{8.7a}$$

Taking square roots:

$$|\Gamma| = \frac{|U_0|}{|2U_2|}\sqrt{\frac{R_l}{R_{in}}}. \tag{8.7b}$$

If the blocking resistances are equal, i.e. $R_l = R_{in}$, then the term under the root sign is equal to 1 and so

$$|\Gamma| = \frac{|U_0|}{|2U_2|}\sqrt{\frac{P_0}{P_2}}. \tag{8.8a}$$

The signs for absolute values were used because in the general case both U_0 and U_2 may be complex, in which case Γ is also complex. With this in mind, we may change from $|\Gamma|$ to the complex $\Gamma(p)$ as function of the complex frequency, and omit the signs of absolute value:

$$\Gamma(p) = \frac{U_0}{2U_2}. \tag{8.8b}$$

If the arcus (polar angle) of the complex transfer coefficient Γ is φ then, on the basis of the relationship $\Gamma = |\Gamma| e^{j\varphi}$ the two types (decimal and natural) of logarithms of Γ will be obtained in the following manner:

$$\text{decimal:}\ \log \Gamma = \log |\Gamma| + j\varphi^{\log e}, \tag{8.9a}$$

$$\text{natural:}\quad \ln \Gamma = \ln |\Gamma| + j\varphi. \tag{8.9b}$$

The attenuation of a filter is a value which characterises the relationship between the power fed into it and that passed through or suppressed by it. It may be explained by saying that its value in the blocking range should be high and in the pass band should be low. It is advisable to measure attenuation in logarithmic units, although it can also be given in nepers and decibels, or even as the relation between powers and voltages. Before examining the correlation between attenuation and the transfer factor Γ, let us express the attenuation (or damping) as the power ratio, both in nepers and decibels, as the definition of Γ is likewise based on the power ratio.

Attenuation in nepers, with powers is given by:

$$a^{\mathrm{Np}} = \frac{1}{2} \ln \frac{P_0}{P_2} \tag{8.10a}$$

and in decibels, again with powers, by

$$a^{\mathrm{dB}} = 10 \log \frac{P_0}{P_2}. \tag{8.10b}$$

Let us now take the two types of logarithm of (8.8a). In logarithms to base 10 both sides of the equation are multiplied by 10:

$$\ln |\Gamma| = \ln \left| \frac{U_0}{2U_2} \right| = \frac{1}{2} \ln \frac{P_0}{P_2}, \tag{8.11a}$$

$$20 \log |\Gamma| = 20 \log \left| \frac{U_0}{2U_2} \right| = 10 \log \frac{P_0}{P_2}. \tag{8.11b}$$

Comparing (8.11) with (8.10)

$$a^{\mathrm{Np}} = \ln |\Gamma| = \ln \left| \frac{U_0}{2U_2} \right|, \tag{8.12a}$$

$$a^{\mathrm{dB}} = 20 \log |\Gamma| = 20 \log \left| \frac{U_0}{2U_2} \right|. \tag{8.12b}$$

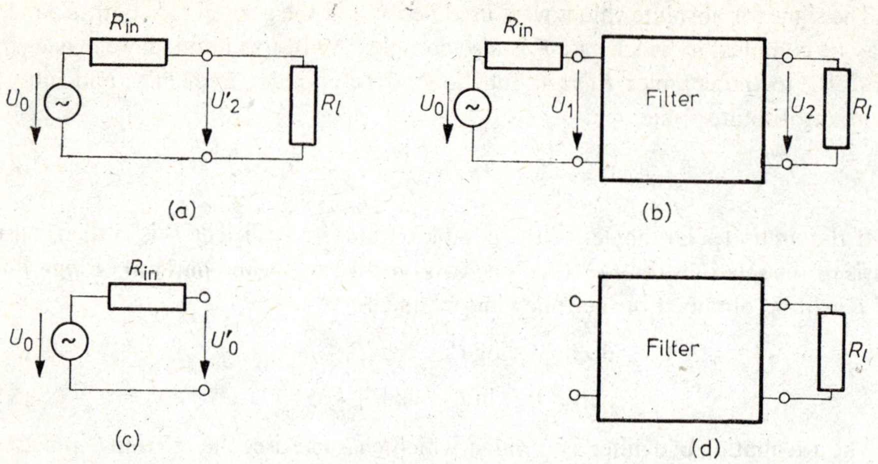

Fig. 8.3. *Definition of the insertion and connection attenuation of filters:* (a) *filterless circuit,* (b) *filter inserted,* (c) *and* (d) *connecting a filter blocked by a load resistance to the idling voltage source*

This proves that, from the absolute value of the complex transfer factor Γ, attenuation values can be directly calculated in both nepers or decibels. Equations (8.12) give the so-called working attenuation of the filter but they apply only if the criterion $R_l = R_{in}$ is met.

In our further investigations we shall deal only with the characteristics which are expressed in decibels. It has been assumed that $R_l = R_{in}$. This means that in a filterless circuit (Fig. 8.3) there is power matching, and on the load a voltage of U_2' is applied. If the filter is inserted into the circuit (Fig. 8.3b) the matching will end, because the primary side input resistance of the filter is blocked on the secondary side by R_l, so this primary side input resistance seldom agrees with R_l, not even with $R_l = R_{in}$. Without matching insertion attenuation takes place which, in terms of decibels, assumes the form

$$a_{\text{insert}}^{\text{dB}} = 20 \log \frac{U_2'}{U_2} \tag{8.13}$$

and the decibel value must be calculated from the quotient of the output voltages before and after the moment when the filter is inserted.

It is worth noting that, if $R_l = R_{in}$, the insertion values and the values of the working attenuation according to (8.12b) will be equal. When inserting the filters, not only R_l and R_{in} be equal but they must also be equal to the input and output impedances of the filter, because this is required to ensure the matching that existed before the insertion of the filter.

In relation to the use of filters, another attenuation can be defined. It is the interfacing attenuation (Figs 8.3c and d) which arises when the filter loaded by R_l (Fig. 83d) is connected to the idling voltage U_0' and the output voltage U_2 on the filter output (provided that the two systems are interconnected), U_2 being equal to the value used in the computation of the insertion attenuation. Accordingly, the connection attenuation is

$$a^{dB}_{connection} = 20 \log \frac{U_0'}{U_2}. \tag{8.14}$$

If the filter is matched to the voltage source, i.e. if $R_{in}=R_{inlet}$ (cf. Fig. 8.2) then, according to the voltage division (since 50% of $U_1=0.5\ U_0$ and since $1/2=-6$ dB), we must have

$$a^{dB} = a^{dB}_{insert} = a^{dB}_{connection} - 6\,\mathrm{dB}. \tag{8.15}$$

Of the three attenuations, the manufacturers of filters usually quote only one, for the suitable correction of the filtered component.

Having treated the matching, transfer, power suppression and attenuation properties of filters, let us now treat their frequency properties.

Characteristics and frequency properties of filters

Filter characteristics are determined by the range of their frequency transfer, i.e. the frequency response of their attenuation. Accordingly, there are

— low-pass (high cut-off),
— high-pass (low cut-off),
— band-pass,
— band-rejecting filters.

The attenuation of filters in the pass band must be below, and in the cut-off band above, a specified value. Filters are usually produced by connecting reactive (C and L) circuit elements (fundamental members) in series but sometimes resistive-capacitive (R and C) elements. (R – resistive, C – capacitive, L – inductive and L is the symbol of the self-induction coefficient.) Since the filters having the steep characteristics required for measurements embody mainly C and L elements, we shall confine our examinations to them. Each basic filter member has a "T" and "π" variant (the designation refers to the structure of the fundamental member, Fig. 8.4). The number of series-connected filter members is determined by the accuracy required and the slope of the cut-off.

The typical basic filter members are shown in Fig. 8.4 together with their T and π variants and main characteristics. For clarity the relative voltage transfer characteristics (the inverses of the corresponding attenuation characteristics) have been indicated, not the attenuations themselves.

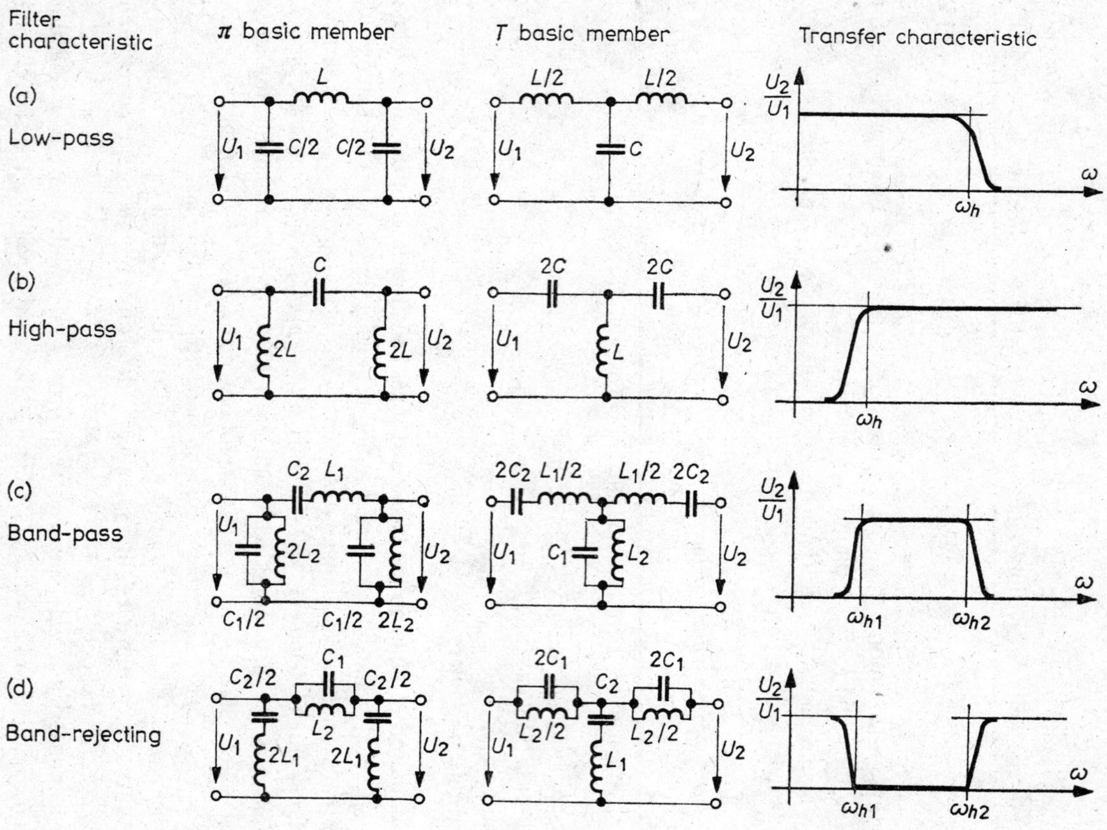

Fig. 8.4. *Characteristic reactant basic filter members and their transfer characteristics*

Low and high-pass filters

The impedance of the coil having self-induction coefficient L (coil) — $Z_L = j\omega L$ — at high frequencies is high, while the impedance of the capacitor having a capacitance C $Z_C = \frac{1}{j\omega C}$ at the same frequencies is either that of good conductor, or close to zero. Therefore, low-pass filters at high frequencies behave nearly like short circuits but they let low frequencies pass without any difficulty. In a similar way, in high-pass filters inductance in the transversal branch at low frequencies is shorted, while at high frequencies their resistance increases with the frequency. The curve of the transfer function of an unattenuated filter (shorted on its output) is relatively steep.

The limit frequency is the resonance frequency of the LC member (oscillating circuit) constituting the filter member, which can be determined using the well known Thomson formula:

$$\omega_l = \frac{1}{\sqrt{LC}} \quad \left(f_l = \frac{\omega_l}{2\pi}\right). \tag{8.16}$$

To prevent self-oscillation in the inputs and outputs of an LC filter, blocking resistances of size

$$R = \sqrt{\frac{L}{C}} \tag{8.17}$$

are necessary. These, however, affect the slope of the transfer characteristic. Since the output resistance of the amplifier and the input resistance of the indicator are generally

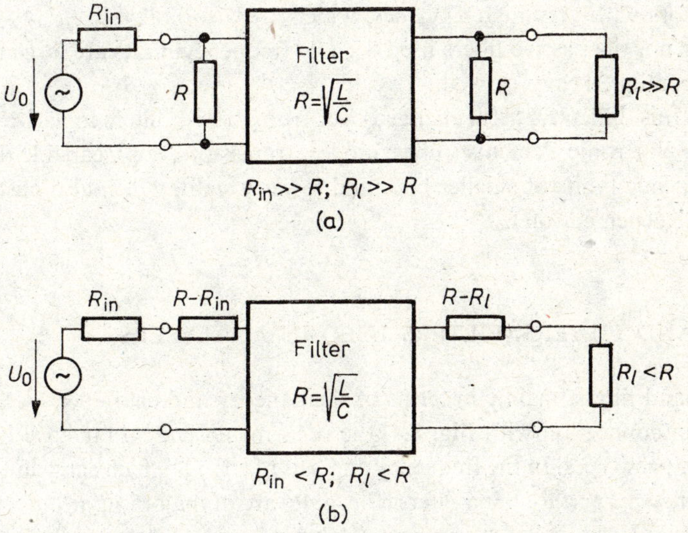

Fig. 8.5. *Insertion of a filter between the signal source (amplifier) and the load (indicator):* (a) *with parallel resistances* (b) *with series resistances*

different from the values in Eq. (8.17), matching must be performed with either a parallel or a series-connected resistance (Fig. 8.5).

The blocking resistances in (8.17) correspond to a matching attenuation of 3 dB at the value ω_l of the frequency limit, in relation to the basic attenuation measurable with the medium frequencies of the band-pass range.

Filter manufacturers generally specify the transfer and attenuation characteristics of their filters. The original slope characteristic of a filter can be made steeper by connecting several filter members in series. In that case, however, the matching must be executed with care to prevent undesirable reflection phenomena which might increase attenuation to an excessive degree in the pass band.

As shown in (8.9) there is a frequency-dependent phase shift of $\varphi \cdot \log e = 0.4342\varphi$ between the input and output of the filter. If a low or a high-pass filter is used simply for the separation of two frequency components (e.g. noise suppression, the separation of the self-oscillations of the measured signal and the measuring system, etc.) such a phase shift may be disregarded. If, however, part of a continuous frequency spectrum is filtered out and the residual signal is examined, a phase shift may cause a displacement of certain components and yield false peak values. To avoid such occurrences, distortion-free filters must be used which ensure a 0 or α=constant frequency-independent phase shift in relation to every component. These can then be taken into consideration, with a constant correction value.

As has been mentioned, the frequency limits of filters made up of *LC* elements are given by the Thomson formula (8.16). It shows that the achievement in a circuit of very low frequency limits requires very high inductances and capacitances and very high *L* and *C* values. To achieve high inductances, bulky iron-core coils are necessary and capacitors with very high *C* values, which are also bulky.

Summing up, the passive filters used for low frequency limits are large and, due to their iron content, heavy as well.

To avoid this difficulty, users changed over from passive filters to active filters built up of active electronic elements (tubes, diodes, transistors, etc.), capable of operating at low frequency limits of smaller bulk and lower weight, which also ensured higher selectivity (steeper cut-off).

8.3.2 BAND FILTERS, FILTERS USED IN ANALYSERS

Although band filters, mainly in terms of their theory and characteristics, have been dealt with in connection with Fig. 8.4, the ones mentioned so far have been mainly low- or high-pass types. In the frequency analysis of signals which carry information in their frequency properties, band filters (Fig. 8.4c) are of major importance. The means of frequency analysis, called "frequency analysers" or, in brief, "analysers", are built up of such low- or high-pass filters in series. We shall now survey them, with stress on their practical applications. As Figure 8.4c shows, the "band-pass filter",

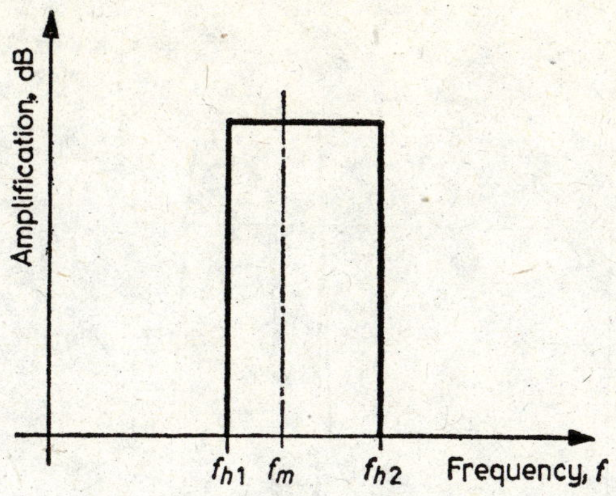

Fig. 8.6. *Ideal filter characteristic (amplification vs. frequency))*

or simply "band filter" passes signals between the boundary angular frequencies ω_{l1} and ω_{l2}. According to the known relationship $\omega=2\pi f$, these angular frequencies correspond to the limit frequencies f_{l1} and f_{l2}. Figure 8.6 illustrates an ideal band filter characteristic.

Since, however, such an ideal characteristic cannot be achieved, we have to try to approximate it with an actual one. Using the notation of Fig. 8.6, let us clarify two important ideas which we will need later on.

The bandwidth of an ideal filter is equal to the difference between the two frequency limits of the band pass:

$$B = f_{l2} - f_{l1}, \quad \text{Hz.} \tag{8.18}$$

To characterise the band filter, a frequency denoted by f_m (mean frequency) is used. It is the geometric mean of the band limits:

$$f_m = \sqrt{f_{l1} f_{l2}}, \quad \text{Hz.} \tag{8.19}$$

A typical variant of the actual characteristic, which best approximates the ideal one in a certain frequency range, is shown in Fig. 8.7. As can be seen, its sides are not vertical and in the pass-band the characteristic is not bounded by a horizontal line, as is the ideal one, but is wavy. The bandwidth is defined by the values at the intersections of its sides with the horizontal line – 3 dB (Fig. 8.7).

(Note that this interpretation of frequency transfer is widely used in transmission engineering, linear amplification, etc.). The bandwidth thus defined (in Fig. 8.7) is also called the 3 dB bandwidth.

In addition to the -3 dB bandwidth another so-called effective bandwidth is also defined. It is related to the response of filters to noise, and assumes its greatest importance in the analysis of signals which carry non-deterministic information. By

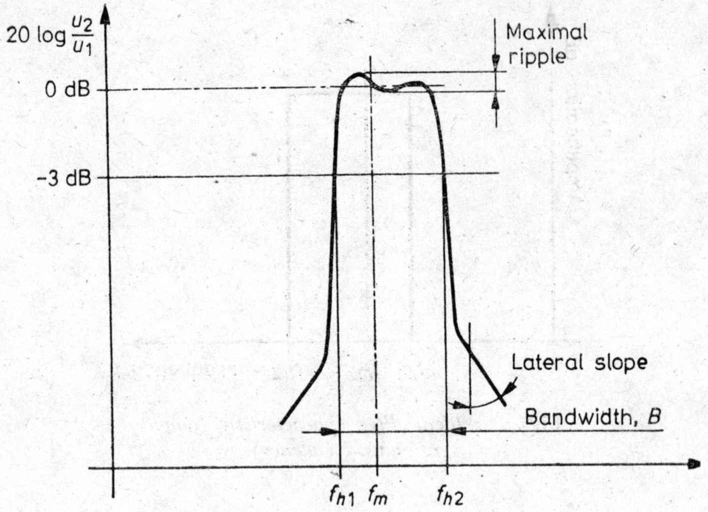

Fig. 8.7. *Real band filter characteristic and the definition of the — 3 dB bandwidth*

definition, the effective bandwidth is equal to the bandwidth of an ideal filter of the same pass as that of the filter examined in the case of white noise.*

The effective bandwidth can be determined by plotting the amplitude response of the filter on a linear scale (not in dB). Noise being the quantity concerned, we seek the square of the curve. (Since the relative values are less than unity, that curve is below the curve of the amplitude response throughout (Fig. 8.8)).

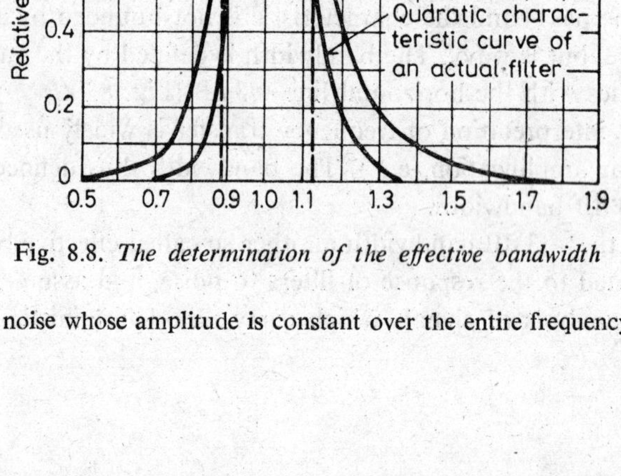

Fig. 8.8. *The determination of the effective bandwidth*

* White noise is noise whose amplitude is constant over the entire frequency range.

Determining the area below the curve of the quadratic function (integrating the function), the effective bandwidth sought is obtained when the area is divided by the highest value of the ordinate (Fig. 8.8).

Different bandwidth definitions for the band filters of analysers

As has been seen, the width of an ideal band filter was given by (8.18).

If the filter characteristics are known, the – 3 dB bandwidth of the effective band filter can be uniquely defined (Fig. 8.7).

Frequency analysers cover a wide frequency range with filters having characteristics close to one another (filter sets). The filters in a set may have

— constant absolute, or
— constant relative bandwidths.

In the case of constant absolute bandwidth, for each filter in the set we have

$$B = \Delta f = f_{l2} - f_{l1} = \text{const.} \quad (8.20)$$

This means that the set of filters (the analyser) covers the full frequency range with a series of filters with constant (e.g. 100 Hz) bandwidth. If the characteristics of the filters having constant absolute bandwidths are plotted on a linear or logarithmic frequency scale, we obtain Fig. 8.9 (for simplicity, the figure shows the characteristics of ideal filters).

In the case of constant relative bandwidth, for each filter in the set we have

$$\frac{B}{f_m} = \frac{\Delta f}{f_m} = \frac{f_{l2} - f_{l1}}{\sqrt{f_{l1} f_{l2}}} = \text{const.} \quad (8.21)$$

so

$$B = \frac{n f_m}{100}, \quad (8.22)$$

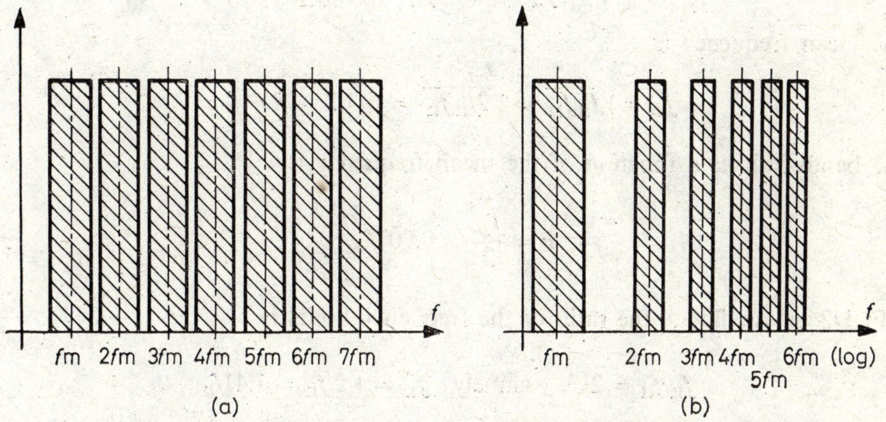

Fig. 8.9. *Band filters (analysers) with constant absolute bandwidth:* (a) *linear frequency scale,* (b) *logarithmic frequency scale*

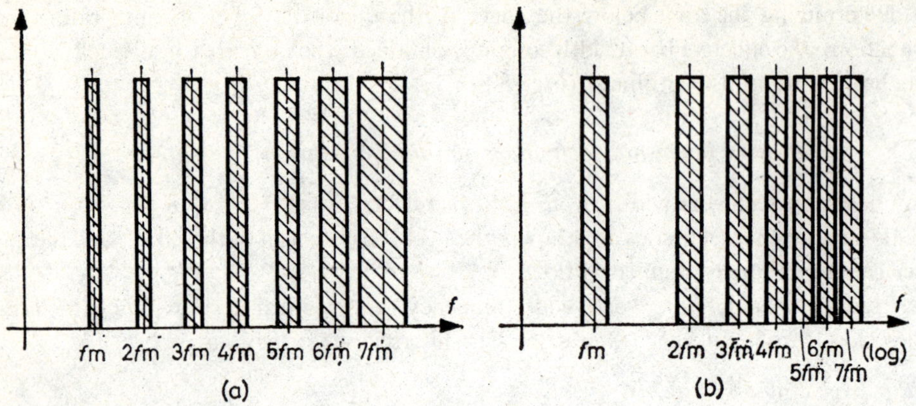

Fig. 8.10. *Band filters (analysers) with constant relative bandwidth:* (a) *linear frequency scale,* (b) *logarithmic frequency scale*

i.e. the bandwidth is equal to n% of the mean frequency f_m. Again for the sake of simplicity, filters of constant relative bandwidths (with ideal characteristics) are shown in Fig. 8.10.

In frequency analysis, octave, 1/2 octave and 1/3 octave (known also as third) filters are used. They all belong to the group of filters with relative bandwidths.

On the basis of the definition of the bandwidth, further significant information can be obtained from filters and analysers made up of filters.

Before going into detail, we draw the reader's attention to Chapter 5 which provided information about the relative units of frequency.

In octave filters, the octave being the logarithm to base 2 of the frequency ratio, the bandwidth is ($f_{l2}/f_{l1}=2$ that is: $f_{l2}=2f_{l1}$)

$$B = f_{l2}-f_{l1} = 2f_{l1}-f_{l1} = f_{l1}.$$

The mean frequency is

$$f_m = \sqrt{f_{l1}f_{l2}} = \sqrt{2f_{l1}f_{l1}} = f_{l1}\sqrt{2} = 1.41f_{l1}.$$

The bandwidth as a function of the mean frequency is

$$B = \frac{f_m}{2} = 0.707f_m.$$

In 1/2 octave filters, the ratio of the frequency limits is

$$f_{l2}/f_{l1} = 2^{1/2}, \quad \text{namely} \quad f_{l2} = \sqrt{2}\,f_{l1} = 1.41f_{l1}$$

the bandwidth

$$B = f_{l2}-f_{l1} = \sqrt{2}\,f_{l1}-f_{l1} = f_{l1}(\sqrt{2}-1) = 0.41f_{l1}$$

and the mean frequency

$$f_m = \sqrt{f_{l1} f_{l2}} = \sqrt{1.41 f_{l1} f_{l1}} = 1.189 f_{l1}$$

and

$$B = 0.344 f_m.$$

With a 1/3 octave filter (third filter) the ratio of the limit frequencies $f_{l2}/f_{l1} = 2^{1/3}$ i.e.

$$f_{l2} = 2^{1/3} f_{l1} = \sqrt[3]{2}\, f_{l1} = 1.259\,9 f_{l1},$$

the bandwidth

$$B = f_{l2} - f_{l1} = 1.259\,9 f_{l1} - f_{l1} = 0.259\,9 f_{l1},$$

the mean frequency

$$f_m = \sqrt{f_{l1} f_{l2}} = f_{l1}\sqrt{2^{1/3}} = f_{l1} 2^{1/6} = 1.122 f_{l1},$$

and

$$B = 0.231 f_m.$$

If the bandwidths are expressed in terms of % of the mean frequencies, then the bandwidth of the octave filter is 70.7%, that of the 1/2 octave filter is 34.4%, and that of the third filter is 23.1%.

Figures 8.11 and 8.12 illustrate the characteristics of the filter sets of the Brüel and Kjaer Co. which are extensively used (in acoustics and in oscillation and noise measurements, etc.) Figure 8.11 shows their 1613 type octave filter, and Fig. 8.12 their 1616 type third filter [6].

This survey of the means used for the analysis of signal frequencies would be incomplete without also treating the latest digital analysers. Digital frequency analysis is based on the digitalisation of the analogue signal examined by resolving its conti-

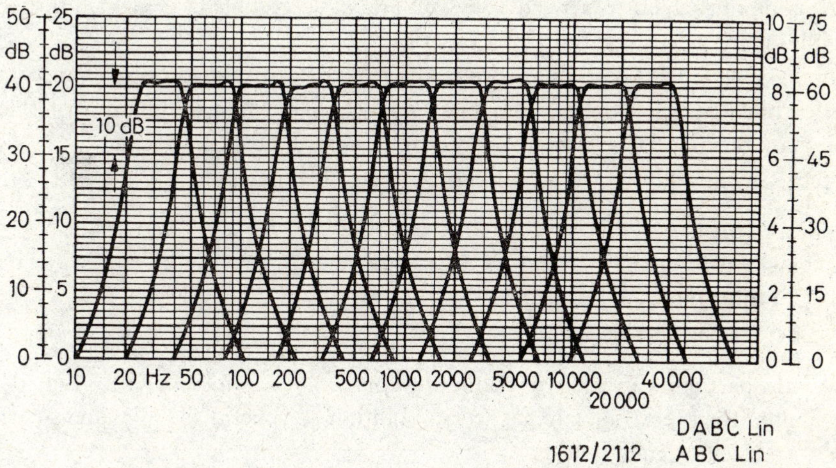

Fig. 8.11. *Characteristics of the filter sets of a frequency analyser with octave filters (Brüel and Kjaer 1613)*

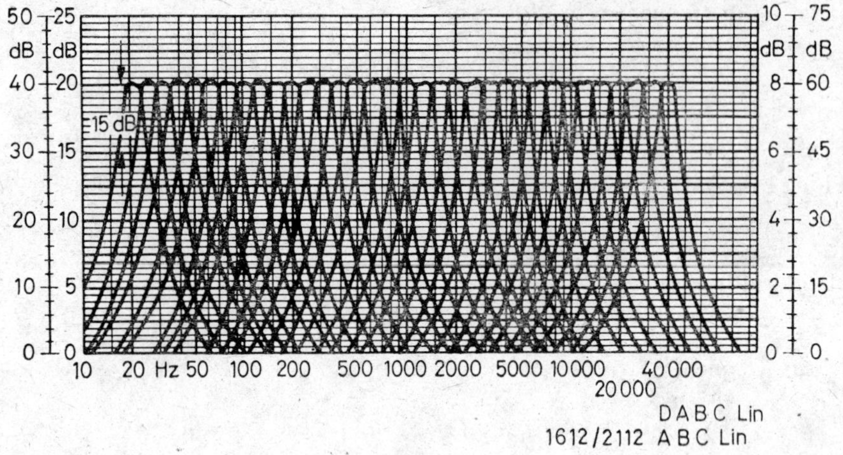

Fig. 8.12. *Characteristics of the filter sets of a frequency analyser with third filters (Brüel and Kjaer 1616)*

nuity in time by sampling and its continuity in amplitude by quantization. Each of the discrete functional points thus obtained represents a digital word, and the words, or more precisely the pieces of digital information they carry, are electronically processed. In principle, groups of digital characters are easy to store, transmit, transform to other codes or change back into analogue signal points.

In the signal resolution a large number of data must be handled. The only way to solve this task is to use computing methods.

To use a large computer for similar jobs proved to be cumbersome. It was the appearance of microprocessors and microcomputers on the market which enabled the construction of such devices in suitable sizes and at acceptable price, which turned the principle of frequency analysis into practice.

Making use of the possibilities offered by microelectronics numerous instrument factories build and sell digital frequency analysers for the examination of the frequency composition of various periodic or non-periodic phenomena.

Typical signal analysers are built up of the following units:

— processor,
— signal storing unit,
— A/D transducers,
— different peripherals.

From the part units a variety of configurations can be assembled, with a high degree of flexibility, for special jobs (e.g. the examination of general signal shapes, periodic signals, transient signals, etc.).

A brief description of the more important units and their functions is given in the following.

Processor unit

To meet the tasks of the processor mostly microprocessors are used, sometimes supplemented with a screen. This latter allows the plotting of a raster (for scaling), based on the arithmetical capabilities of microprocessors. Depending on the type of the microprocessor the processor can process data or function points (coordinates) transformed into 8—10-bit code words.

For practical reasons the processors of such signal analysers differ from those applied in general computer operations. In contrast to the general purpose devices which mostly switch over to the decimal system, they not only apply the 8—10-bit code words without transformation, but also keep the result in binary form.

Signal storing unit

The coded digital signal produced by the A/D coverter (digitaliser), viz. a set of code words from the analog signal to be analysed, are received and stored in the memory of the signal shape storing unit. It is mostly this unit which houses the drives of the system's peripherals, the interfacing units, the controls of the manipulating bus, etc. The device can generally pick up the signals on several channels.

A/D converter

The available A/D converters operate on different principles. We do not wish to go into details on these principles because they vary from one manufacturer to the other. However, irrespective of the type, they are generally programmable sampling digitalisers, and common types, depending on whether the signal to be analysed is periodic or aperiodic.

Sampling digitalizers can only be applied for periodic signals. The samples picked up, stored and then set back in the required sequence, produce a figure which is identical to the signal shape examined. The fundamental difference from the digitalization of transient signals is that while the function points of a transient signal originate from the same signal, those produced by a sampling digitalizer come from newer and newer periods of the signal under study.

The other digitalizer takes samples in a continuous process from the analogue signals sent – similarly continuously – to the input, and converts the samples into binary code words at the same rate. The code words appear on the output in a "bit parallel/byte serial" sequence.

Peripherals

Peripherals are applied according to needs; they may be printers, X–Y recorders. If a general purpose device having no built-in monitor is used as processor, it is advisable to apply a screen, which displays both graphic and alphanumerical information.

Summary

The manufacturers' instructions provide detailed information on the possible configurations of the system (the operation of the units, the resolution of the time and the amplitude of the signal, the accuracy of the resolution, etc.).

8.4 THE OBJECTIVES OF FREQUENCY ANALYSIS

We have now surveyed the basic tasks and the most important methods of frequency analysis. In this Section we shall present some details.

8.4.1 TYPES OF ANALYSED SIGNALS

As we have mentioned, in frequency analysis not only the highest value of the frequency signal carrying the measurement information is measured, but also the amplitude level of each individual frequency component.

Here is a brief survey of the signal types which carry measurement information in their frequency properties. We refer back to Chapter 3 which gives a short summary of time functions and signals.

In frequency analysis the following signals are of importance:

— harmonic,
— periodic,
— aperiodic, and
— irregular signals (noise).

With a harmonic signal, the only problem is how to ascertain whether it contains only one harmonic or more. As an approximation, a display on the oscilloscope screen may be acceptable, while in dubious cases an accelerated analysis by analyser may give the correct answer.

A periodic signal consists of a finite number of components with discrete frequencies. Accordingly, the spectrum may be as shown in Fig. 8.13 i.e. as a line or a discrete spectrum. The spectrum is plotted by marking the discrete frequencies on the frequency axis in the amplitude *vs.* frequency coordinate system, and then measuring the amplitude of the harmonic component of the frequency concerned. (According to the basic interrelations of the Fourier series, these discrete frequencies are integral multiples of one another, and of the basic frequencies.

If aperiodic signals are examined (e.g. in shock analysis) the signal has but a single run and is not repeated because the physical phenomenon which produces it (e.g. excitation, force, potential, etc.) becomes stabilised and remains unchanged. Figure 8.14, for instance, shows an aperiodic signal and the corresponding characteristic

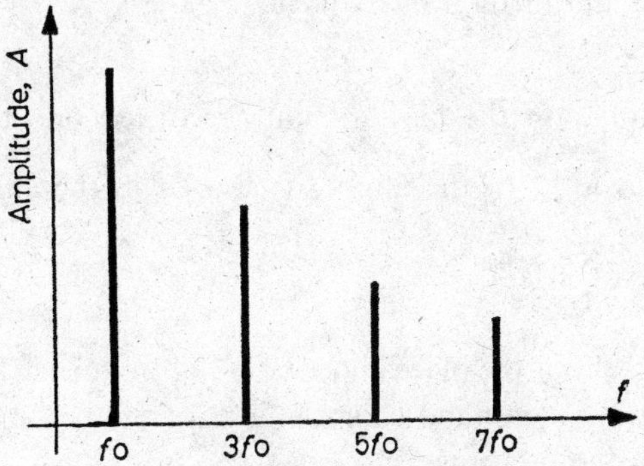

Fig. 8.13. *Example of a line spectrum*

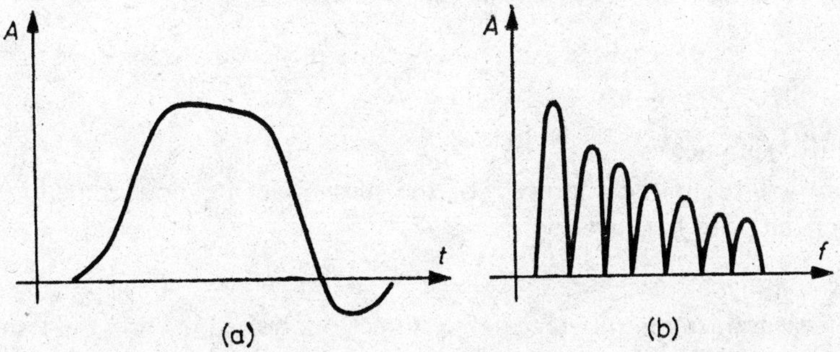

Fig. 8.14. *An aperiodic signal:* (a) *its time function,* (b) *its frequency spectrum*

spectrum. Such aperiodic signals, also called transient signals, come about when a process is started, or when a change in one or several previously constant characteristics of a process (which can be described by constant or variable characteristics) occurs (e.g. on a collision).

The spectrum in Fig. 8.14 consists of a large but finite number of sections which cover the entire frequency range. The case in Fig. 8.14 is rather unusual but possible.

In the analysis of irregular signals (noise), the frequency spectrum is continuous i.e. the signals are often limited in frequency band and thus the beginning and the end of the spectrum are at discrete-finite values. For more about the problems arising with noise-type signals, see Chapter 9.

8.4.2 PRACTICAL PROBLEMS IN FREQUENCY ANALYSIS

Before starting frequency analysis, it is important to clarify some problems concerning the quantity of the object to be measured, the type of the signal to be analysed for frequency, and the amount and accuracy etc. of the information to be obtained.

First of all, we have to decide

— about the place and method of frequency analysis;

(a) on-site,
(b) in the laboratory, on the basis of data recorded on-site on magnetic tape;

— about the bandwidths of the filters to be used in the analysis:

(a) octave, third, etc.
(b) narrow-band analysers.

To reach a decision about the place and method of frequency analysis, the best way is to start from the advantages and shortcomings of the two processes — i.e. direct and indirect (intermediate) storage.

Direct frequency analysis on site

In Section 8.1. reference was made to direct frequency analysis as one of the possibilities. If offers the following advantages:

(a) relative simplicity,
(b) repeatability: if the character of the measurement allows repetition, major errors etc. can be corrected.

Its shortcomings:
If the measurement is not repeatable, direct analysis does not yield sufficiently reliable results. Subjective errors and external disturbing events may cause loss of measurement information.

Frequency analysis in the laboratory and intermediate storage on magnetic tape

The second possibility for frequency analysis is recording the measurement information on magnetic tape on site, and processing it in a laboratory.

Advantages:

(a) The method eliminates the shortcoming of non-repeatability and the ensuing uncertainty. Recording some non-repeatable information makes it repeatable.

(b) By making records of each frequency band, non-stationary phenomena (fluctuating revolution speed, slightly changing frequency composition etc.) can be eliminated during playback.

(c) It allows frequency tranformation i.e. by slow (fast) recording and fast (slow) playback, the frequency bands can be transposed, i.e. transferred to different ranges. By this method low-frequency (e.g. mechanical) processes can be analysed even if only high-frequency filters are available (a frequent case), and vice versa.

(d) There is no need to set up sensitive, heavy and bulky analysers on-site.

Disadvantage:

If the measurement can be repeated, direct analysis provides sufficient reliability but if it is repeated on the measured object and not on a magnetic tape, the errors will be easier to filter out.

The following two block diagrams illustrate the methods and apparatus used in direct and intermediate storage frequency analysis.

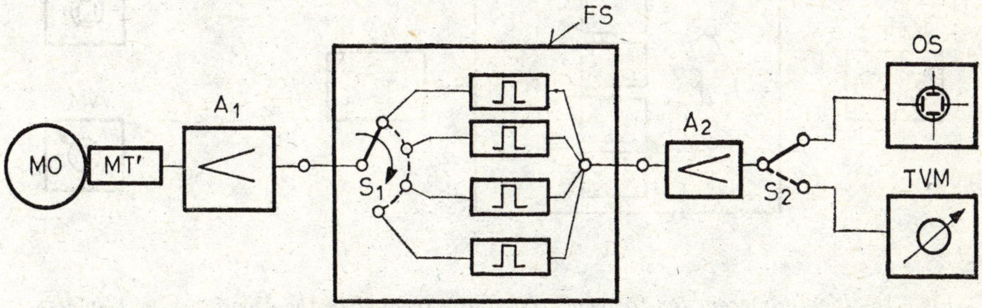

Fig. 8.15. *Block diagram for direct on-site measurement and frequency analysis:* MO — *the measurand object,* MT — *measuring transducer,* A_1, A_2 — *amplifiers,* FS — *filter set,* OS — *oscilloscope,* TVM — *tube voltmeter (levelmeter),* S_1 — *switch for selecting the frequency band,* S_2 — *switch for selecting the indicator*

Figure 8.15 shows the basic arrangement for direct measurements on-site, and frequency analysis. After amplification and signal shaping, the electrical signal obtained at the output of the measuring transducer reaches the filter set where the filtering frequencies, and the bands to be examined, can be set by means of the selector switch S_1. The frequency components, described in terms of the nominal filter frequencies, can be directly read from the electronic tube voltmeter that is level meter in dB. It is advisable to insert the oscilloscope shown in the figure into the circuit, because it shows the shape of the signal, and signal components, remaining in the bands after filtration.

Figure 8.16 shows two block diagrams for frequency analysis executed in the laboratory, with intermediate magnetic tape storage. Figure 8.16a is the block diagram for on-site measurement and (b) is that for frequency analysis (in the laboratory, using the stored data).

Measurement or frequency analysis is executed as shown in Fig. 8.15. However, measurement and analysis are separated in space and time by the tape recorder, since the data it records can be analysed away from the site, and at another time.

The next important practical problem is to decide whether filters of constant absolute or constant relative bandwidth should be used in the analysis. We shall mention some points for consideration, referring to Figs 8.9 and 8.10.

Filters of constant absolute bandwidth should be used in the following cases:

1. In frequency analysis at low frequencies.

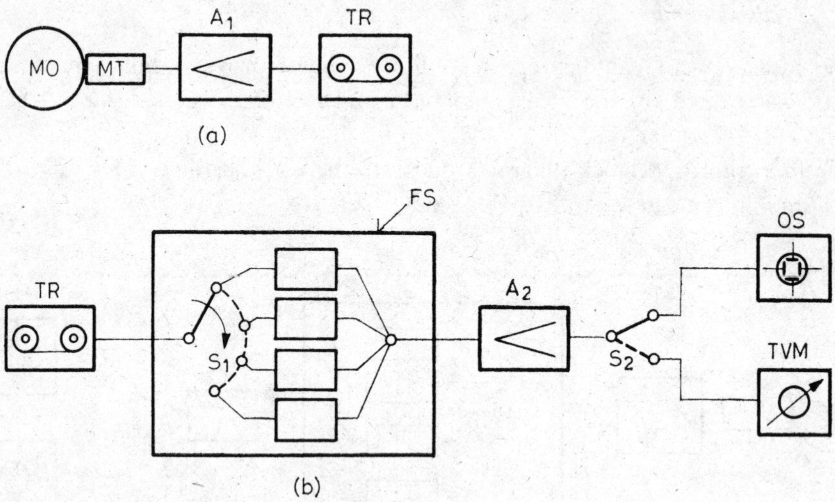

Fig.8.16. *Block diagram for frequency analysis executed in the laboratory with intermediate magnetic tape storage:* (a) *block diagram for on-site measurement,* (b) *block diagram for frequency analysis in the laboratory,* MO — *measurand object,* MT — *measuring transducer,* A_1, A_2 — *amplifiers,* TR — *tape recorder,* FS — *filter set,* OS — *oscilloscope,* TVM — *electronic tube voltmeter,* S_1 — *switch for selecting the frequency band,* S_2 — *switch for selecting the indicator*

2. In the analysis of a discrete, frequency spectrum or a continuous frequency spectrum in which the maxima are closely spaced. In such cases every individual spectral component, or two components corresponding to closely-spaced maxima, can be filtered out and defined: this cannot be done using constant relative bandwidths. (Note: with an analyser made up of filters of 8.5% relative bandwidth examination cannot go beyond the 10th — or in very advantageous cases — the 11th harmonic component).

Filters of constant relative bandwidths should be used (preferably) in the following cases:

1. In frequency analysis at high frequencies. At low frequencies, the pass band of filters with constant relative bandwidths is very narrow and causes the pointer of the device (e.g. electronic tube voltmeter) connected behind the filter to oscillate, making evaluation difficult).

2. For a continuous frequency spectrum, because high frequency components seldom contribute to the spectrum.

3. In the frequency analysis of insufficiently steady — fluctuating — processes (e.g. fluctuations in rpm., in tape speed, etc.).

4. If the task is to examine the effect of a physical quantity (oscillation, noise, acoustic signals, etc.) upon humans, buildings, machines, etc., frequency analysis with constant relative bandwidths is generally sufficient to meet the case.

The next problem is to decide upon the preferred bandwidths of the filters in the frequency analyser. A definition of the optimum bandwidth can be approached from two viewpoints, technical and economic.

From the technical viewpoint, it is obvious that the amount of information obtainable by analysis is inversely proportional to the bandwidths of the filters. This means that the narrower the bandwidths, the more accurate is the information obtained, and the better the signal examined can be broken down to basic components.

From the viewpoint of economy, we have come to the conclusion that not only is the price of the filters inversely proportional to their bandwidth, but so is the time needed for the examination, and therefore the cost involved. To cover the measuring range with filters of lower bandwidths, we would need more filters and also more time for the analysis, which would naturally entail higher costs.

The result of these technical and economic considerations is as follows:

In practice, the optimal bandwidth is chosen to be only as narrow as to yield sufficient information in the minimum time and at the minimum cost.

Thus in the case of decreasing bandwidth, we distinguish octave, third and narrow-band analysis. Decreasing bandwidths provide increasingly accurate results.

In the example which follows we shall compare the results obtainable by filter sets of different bandwidths. The example will at the same time show the selection of the best bandwidth for the task in hand. It was taken from [17], and refers to methods used in factories making concrete products: it deals with the spectrum of the noise a concrete-compacting skaker table (the noise source). Figure 8.17 shows the various spectra with logarithmic scales (level and frequency). As it is evident from the figure, the largest amount of information is contained in the narrow-band spectrum. With

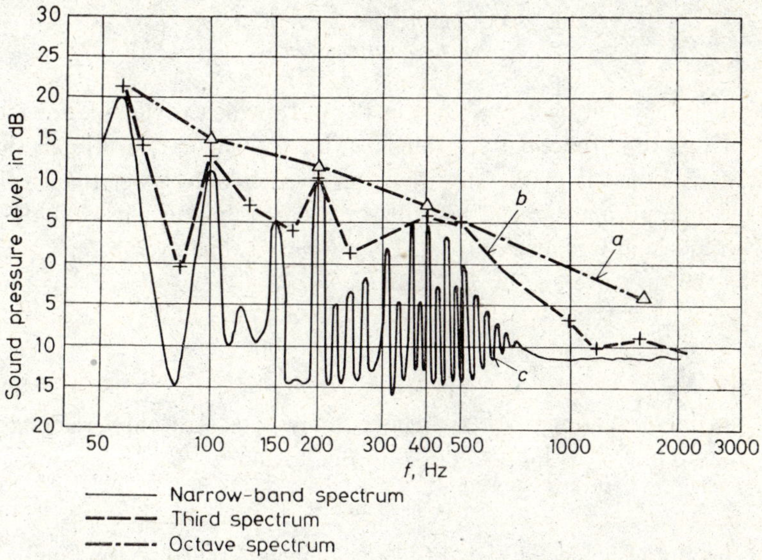

Fig. 8.17. *Spectrum of the sound pressure of a noise source [17]*

increasing bandwidth, the information content decreases. In many cases the octave spectrum provides sufficient information about the spectral composition, in a minimum of time and at a negligible cost.

The third spectrum throws light upon certain peculiarities. The determination of the third spectrum is necessary for the study of the effects of the spectral components on machines or equipment. If the frequency spectrum must be examined in great detail, narrow-band analysis must be executed, with so-called narrow-band analysers.

The octave and third analysers are built up from the filter sets described before. In principle narrow-band analysers can be built up in the same way, but for narrow-band analysis some other devices are preferable, some of which will be the subject of the next section.

8.4.3 NARROW-BAND FREQUENCY ANALYSIS, NARROW-BAND ANALYSERS

If there exists an analogue multiplier which can multiply two time functions then, on the basis of the orthogonal properties of trigonometric functions, it will also be suitable for narrow-band frequency analysis.

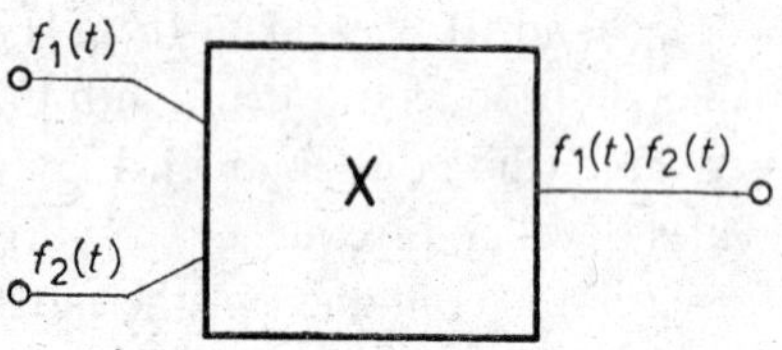

Fig. 8.18. *Analogue multiplier used as a narrow-band analyser*

Let us study the multiplier as shown in Fig. 8.18 which is suitable for the multiplication of two time functions $f_1(t)$ and $f_2(t)$. $f_1(t)$ and $f_2(t)$ are input signals. If its frequency response is of the kind that can transfer only lower frequencies than those of the multiplicand signals, we call it "inert". In this case, it will only average signals with too high frequencies for it, i.e. produce a time average.

Denoting the time average by F_T we have

$$F_T = \frac{1}{T} \int_{t_0}^{t_0+T} f_1(t) f_2(t)\, dt. \tag{8.23}$$

To use an inert multiplier for frequency analysis, a sinusoidal reference signal must be fed into one of its inputs, (e.g. f_1 in Fig. 8.18):

$$f_1(t) = \sin \Omega t, \tag{8.24}$$

into the other input is fed the signal to be examined, $f_2(t)$ (time function). For frequency analysis we examine a component of the signal of frequency ω, that is, its close frequency surroundings. If the signal at frequency ω and close to it assumes the shape of $f_2(t)=A\sin\omega t$ or $f_2(t)$ then at the frequency of ω its Fourier series will have a single harmonic component of amplitude A, and with the sinusoidal reference signal (8.24) we arrive at

$$F_T = \frac{1}{T}\int_{t_0}^{t_0+T} A\sin\omega t\sin\Omega t\,\mathrm{d}t. \tag{8.25}$$

If $\Delta\Omega$ is the difference between the examination frequency ω and the reference frequency Ω then

$$\omega = \Omega+\Delta\Omega \tag{8.26}$$

and

$$F_T = \frac{1}{T}\int_{t_0}^{t_0+T} A\sin(\Omega+\Delta\Omega)t\sin\Omega t\,\mathrm{d}t =$$

$$= \frac{1}{T}\int_{t_3}^{t_0+T} \frac{A}{2}[\cos\Delta\Omega t-\cos(2\Omega+\Delta\Omega)t]\,\mathrm{d}t. \tag{8.27}$$

As regards the analogue multiplier, two types are considered here, notably

— the electrodynamic variant and
— the Hall generator.

The electrodynamic instrument may be made without iron (having an air core) or be ferrodynamic, i.e. with an iron core. Figure 8.19 is a schematic illustration of the measuring instrument with an air core electrodynamic device.

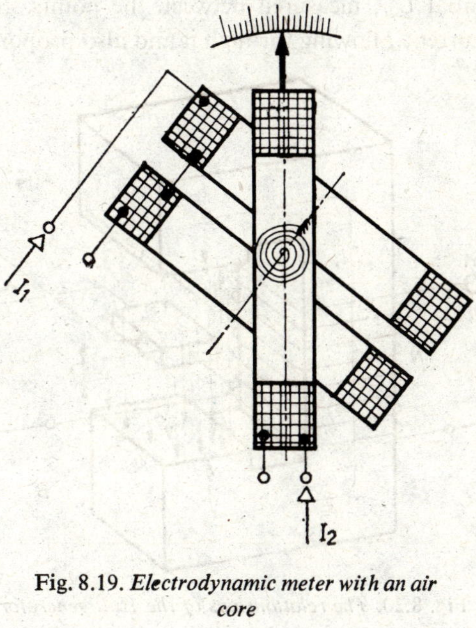

Fig. 8.19. *Electrodynamic meter with an air core*

The angular displacement φ_e of the pointer (rotating coil) of the meter is proportional to the current in the two coils, namely

$$\varphi_e = KI_1 I_2, \tag{8.28}$$

i.e. the angular displacement of the pointer is proportional to the product of the currents flowing in the stationary and rotating coils.

Induction devices can also be used to obtain products. They work on the principle of the induction motor with squirrel-cage rotors. The currents flowing in the two opposite coils produce two fluxes with a phase shift of β, causing a torque to act on the solid aluminium or copper disk in the resultant electro-magnetic field:

$$M = KI_1 I_2 \sin \beta. \tag{8.29}$$

If the countertorque, which depends on the angular displacement, is linear then the result is similar to (8.28).

The Hall generator operates on the basis of the effect discovered by the U. S. physicist Hall in 1879. The Hall effect is a disturbance of the lines of current flow in a conductor due to the presence of a magnetic field, leading to an electric potential gradient transverse to the direction of the current flow. In the presence of a magnetic field characterised by the intensity of the magnetic field (magnetic induction) B perpendicular to the current flowing through a prismatic conductor or semiconductor, the Hall potential U_H will arise between its lateral edges.

Instead of going into a detailed physical analysis of the Hall potential, we shall confine ourselves to what we need to know about the Hall generator. Let us start from Fig. 8.20. The Hall potential U_H, measured between the points shown in the figure, is proportional to the current I flowing through it and also proportional to the intensity

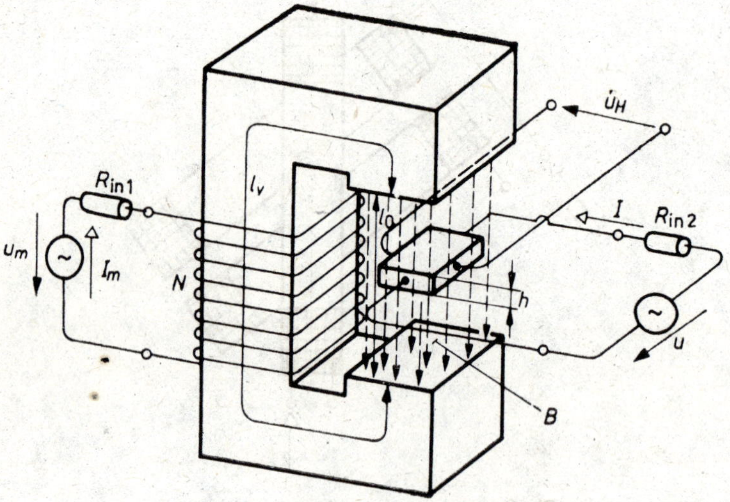

Fig. 8.20. *The relationships of the Hall generator*

of the magnetic field (magnetic induction) B that produces it:

$$U_H = \frac{R_H}{h} IB, \tag{8.30}$$

where R_H is the so-called Hall constant which can be regarded as a material property, not having the dimensions of resistance.

The dimension h can be read from the figure.

Now we wish to know how the Hall generator, as an electrodynamic instrument, can be used for the multiplication of two potentials. Let us excite the magnetic field characterised by the intensity of the magnetic field (magnetic induction) B in the manner shown in the figure, with the coil of pitch N wound on an iron core with an air gap l_0. The length of the iron in the core is l_i.

We may write down the law of excitation for the magnetic circuit:

$$\oint \bar{H}\,\mathrm{d}l = \Sigma I, \tag{8.31}$$

or, with the notation of the figure

$$H_i l_i + H_0 l_0 = N I_m. \tag{8.32}$$

Since, for simplicity, the cross-section of the core may be taken as being identical throughout and, according to the law of conservation of flux, induction is also constant in the whole magnetic circuit:

$$\Phi = \text{const} \tag{8.33}$$

thus

$$B = \frac{\Phi}{A} = \text{const.} \tag{8.34}$$

The values of the magnetic field strength can be expressed in terms of the intensity of the magnetic field (magnetic induction). In the iron

$$H_i = \frac{B}{\mu_0 \mu_i} \tag{8.35}$$

and in the air gap

$$H_0 = \frac{B}{\mu_0}. \tag{8.36}$$

Substituting the latter two equations into (8.32) we have

$$\frac{B l_i}{\mu_0 \mu_i} + \frac{B l_0}{\mu_0} = N I_m \tag{8.37a}$$

whence magnetic induction can be expressed as

$$B = \frac{\mu_0 N}{l_0 + \dfrac{l_i}{\mu_i}} I_m. \tag{8.37}$$

Let us denote the multiplying factor I_m in (8.37) — which is evidently constant — by K_m:

$$K_m = \frac{\mu_0 N}{l_0 + \frac{l_i}{\mu_i}} = \text{const.}, \tag{8.38}$$

which yields a simple linear correlation between magnetic induction and magnetic current:

$$B = K_m I_m. \tag{8.39}$$

Substituting (8.39) into (8.30) and introducing the simplifying notation

$$K_H = \frac{R_H K_m}{h} \tag{8.40}$$

for the Hall potential we get the relationship:

$$U_H = \frac{R_H K_m}{h} I I_m = K_H I I_m, \tag{8.41}$$

which shows that the Hall potential is proportional to the product of the current I flowing across the Hall generator (electrical driving) and the current I_m exciting its magnetic field (magnetic driving).

Passing the Hall potential of the Hall generator to an inert simple permanent magnet with a moving coil known as Deprez instrument, the same situation will occur as with the electrodynamic instrument (cf. 8.28). Accordingly, the readout of the Deprez instrument, when measuring the Hall potential, will be

$$\varphi_H = K'_H I I_m. \tag{8.42}$$

Now let us revert to equation (8.27). For the electrodynamic instrument

$$f_1(t) = I_1(t) = \sin \Omega t,$$
$$f_2(t) = I_2(t) = A \sin \omega t.$$

For the Hall generator:

$$f_1(t) = I(t) = \sin \Omega t,$$
$$f_2(t) = I_m(t) = A \sin \omega t,$$

where $f_1(t)$ is the reference signal and $f_2(t)$ is the signal to be analysed.

Since (8.28) and (8.42) apply, the deflections of the pointer φ_e and φ_H tends toward the time average F_T. (In the formulae it is irrelevant whether an electrodynamic instrument or a Hall generator and attached Deprez instrument are concerned, since we may use the reference signal $\sin \Omega t$ to replace $I_1(t)$ or $I(t)$ and with the examined signal $A \sin \omega t$ to replace $I_2(t)$ or $I_m(t)$.) Therefore we choose a single formula which applies to both cases, i.e. $\varphi_e = \varphi_H = \varphi$.

Due to its inertia, the pointer of the indicator instrument (electrodynamic measuring or Deprez instrument) will tend to settle at the average value F_T. This effort manifest itself in the oscillations of the pointer, caused by the difference between the reference and the examined frequencies. So we may write that

$$\lim_{\Delta\Omega \to 0} = F_T. \tag{8.43}$$

Let us write equation (8.27) again but, in consideration of the boundary value (8.43), replacing F_T by

$$\varphi \cong \frac{1}{T} \int_{t_0}^{t_0+T} \frac{A}{2} [\cos \Delta\Omega t - \cos (2\Omega + \Delta\Omega) t] \, dt. \tag{8.27a}$$

Again due to the inertia of the instrument, the second term under the integral sign can be neglected because its frequency is $2\Omega + \Delta\Omega \gg \Delta\Omega$, whereas the inert instrument responds only to low frequencies ($\Delta\Omega$). In such cases integration is relatively easy:

$$\varphi \cong \frac{1}{T} \int_{t_0}^{t_0+T} \frac{A}{2} \cos \Delta\Omega t \, dt = \frac{A}{2} \left[\frac{\sin \Delta\Omega t}{T\Delta\Omega} \right]_{t_0}^{t_0+T}. \tag{8.44}$$

We have arrived mathematically at the boundary value according to (8.43) which, from the physical aspect, means a closer agreement between the reference frequency Ω and the examined frequency ω; the term in brackets will tend toward 1, as $\Omega \to 0$. Accordingly

$$\varphi \to \frac{A}{2}. \tag{8.45a}$$

In practice, this is equivalent to

$$\varphi \cong \frac{A}{2}. \tag{8.45b}$$

If, therefore, the agreement between Ω and ω becomes closer i.e., if $\Delta\Omega$ decreases, the amplitude of the pointer's oscillation will also decrease and its deflection will be greater. In the boundary case $\Delta\Omega = 0$, oscillation will stop entirely, the approximate equality (8.45) becomes exact and the value can be read off the instrument.

The selectivity of the method (its resolution) is considerable, since great frequency differences $\Delta\Omega$ cause increasing deflection of the pointer to reach a maximum in the case of $\Delta\Omega = 0$. The maximum clearly indicates the frequency and makes the process highly selective.

It is important to remember that this method implies constant absolute bandwidth, i.e., the terms (8.45) derived from the integration of (8.44) are independent of frequency, while with constant relative bandwidth a term linearly dependent on frequency would be inevitable.

In practical cases, depending mostly on the properties of the readout instrument, a 3 to 5 Hz constant absolute bandwidth can be achieved, which approximates to line-type analysis. The examination must take place with frequency changes, to cover the whole range.

8.4.4 REAL-TIME ANALYSIS

Real-time analysis is a modern method of examination. That its theory preceded practical application can be attributed to the fact that it requires sophisticated electronics. Its widespread application began some 15–18 years ago.

As is well known, suitably defined constants (e.g. effective value, absolute mean value, signal power, etc.) facilitate the description of periodic signals. Since the definitions generally contain infinite series, a paradox exists in that the "constants" are constant only if they refer to signals which last for infinitely long time periods. If the examination interval is finite, the "constants" defined according to the classic methods become time-dependent, and represent the only momentary properties of the signal.

Real-time analysis developed from such examinations, executed during finite periods of time. The name "real-time analysis" indicates that the signal examined is not one which was recorded in the past but is a signal actually being examined, and that the examination takes place rapidly — at an accelerated pace — in the real-time of the occurrence of the event.*

Real-time analysis may refer to both the signal level and the frequency of the signal. (For more detail see [8.2].)

We shall deal primarily with the frequency analysis of non-periodic signals, and attempt to outline the most important problems involved. To carry out the task, real-time analysers, now available commercially, are used.

Short-duration or transient signals can be analysed using intermediate magnetic storage, and subsequent repeated playback of the signal from the tape for the examination of each frequency component. This process, however, is not only lengthy but sometimes impossible, since magnetic or other (e.g. digital) recordings of a process cannot always be produced, and the possibility of obtaining results is the use of a real-time analyser.

Real-time analysers and real-time analysis rely on the thesis of "first sampling", from information theory, which states that an aperiodic signal — called band-limited — with a frequency component of f_{max} is completely characterised by the value of its ordinate, observed at intervals of $1/2f_{max}$ or more frequently. If samples of a short duration aperiodic signal are taken with sufficient frequency, on their basis the function to be analysed can be described. "Sufficient frequency" means that the analysis must be executed during a sufficiently short time, i.e. the analysis must take place simultaneously and instantly at each frequency, i.e. within the very short time in which the aperiodic process being studied takes place. Hence the name of the process and device.

*With the appropriate use of sampling theory, if a random signal is fed into an ideal band filter (Fig. 8.6) of bandwidth B, each output from the filter appearing at

*In literature and in professional circles, the process is denoted by its English term Real Time Analysis (abbreviated to RTA); sometimes its German name Echtzeit Analyse is used.

intervals $1/B$ will be statistically independent of the others. Let the duration of the aperiodic signal be T; in this case the number of samples which determine it unambiguously — in other words its degrees of freedom — can be expressed as follows:

$$n = 2BT. \tag{8.46}$$

Naturally, if $B=1/T$, from (8.46) $n=2$.

If, therefore, with a filter of bandwidth B, samplings take place at intervals L/B, the minimum time during which the samples can be collected and averaged for a band of the spectrum of width B will be $n/2B$.

To obtain the full spectrum, the examination takes place between f_1 and f_2 and the measurement must be repeated $(f_2-f_1)/B$ times. This means that the spectrum can be obtained during the time $n(f_2-f_1)/2B^2$. Provided that the data can be processed, the full spectrum will be available during $n/2B$, this being the shortest time for the recording and forming the spectrum.

Although real-time analysis can be done using both analogue and digital methods, the commercially available devices are mostly hybrid. They contain both analogue and digital circuits. The widely used Brüel and Kjaer products [6], upon which we shall concentrate, are of this type.

Time-compression, frequency expansion

Both time-compression and frequency expansion mean the acceleration of the signal. Acceleration reduces the time required for analysis and, if acceleration is sufficient, then the many band filters at the output, with frequency ranges close to one another, will produce a real-time output signal.

The minimum time for recording a full spectrum with n degrees of freedom with a filter of bandwidth B, between f_2 and f_1, is

$$t_{\min} = \frac{n}{2B^2}(f_2-f_1). \tag{8.47}$$

Signal acceleration means that the signal is compressed (it becomes narrower) in the time range, and expands (becomes wider) in the frequency range. Compression and expansion are essentially inversely proportional. In Figures 8.21 and 8.22 a 1/4 time-compression, and the corresponding 4-fold frequency expansion are illustrated.

Compression cuts the duration of the signal and expands the frequency range, i.e. increases the bandwidth required for transfer.

If the signal, prior to analysis, is accelerated by a factor K (compressed), the quantities of time character must be multiplied by $1/K$, and those of frequency character by K. This means that in (8.47) B will be replaced by KB and (f_2-f_1) by $K(f_2-f_1)$. With these changes we get

$$t_{\min} = \frac{nK(f_2-f_1)}{2K^2B^2} = \frac{n(f_2-f_1)}{2KB^2}. \tag{8.48}$$

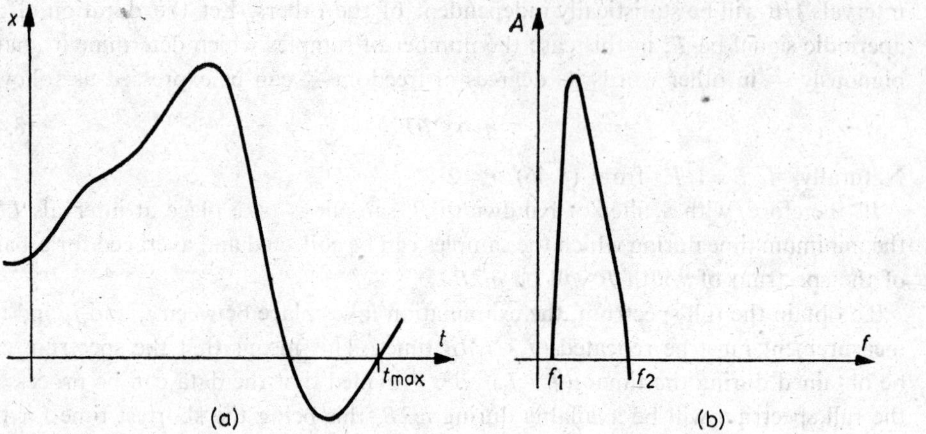

Fig. 8.21. *The original signal:* (a) *time function,* (b) *spectrum*

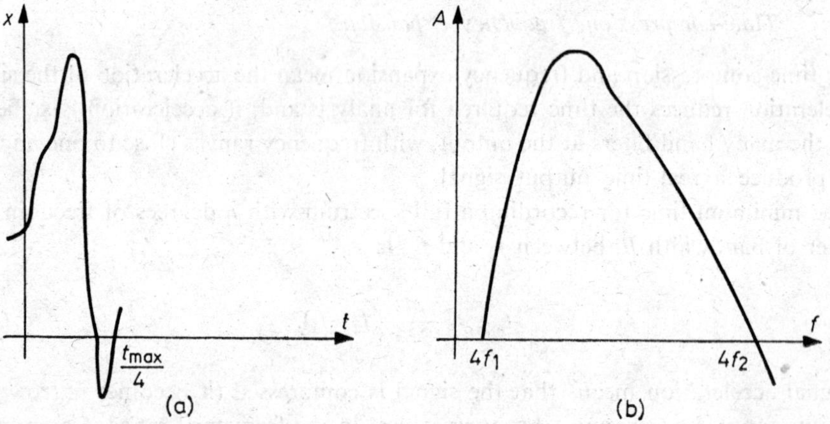

Fig. 8.22. *The time-compressed and the frequency-expanded signal:* (a) *time function,* (b) *spectrum*

Comparison of (8.47) and (8.48) shows that the time required for the analysis has decreased to a Kth part of the original. This means that if K has large values, the real-time analysis of the signal is feasible.

With respect to the degrees of freedom, it should be noted that the use of the resolution capacity β is preferable to that of the bandwidth. Resolution capacity is the distance measured in frequency units between the filters and not the bandwidth. In analysers operating on the time-compression principle, the bandwidth is usually 1.4–1.6 times greater than the resolution.

If in a time compression analyser with degrees of freedom $n=2$

$$2 = 2\beta T, \tag{8.49}$$

then

$$T = 1/\beta. \tag{8.50}$$

Since the frequencies used in practice and called upon for the acceleration of signals the necessary compression coefficient K is great – about 5×10^5 — time compression using tape recording cannot be achieved.

In the 3348 type real-time analyser of the Brüel and Kjaer Co., the signal is accelerated by putting it in a so-called recirculation-type memory, in digital form. The accelerated digital recirculated data are then reconverted into analogue form. This signal is time-compressed and can be analysed by a high-frequency so-called sweep filter. Before the analogue input signal is put into the memory it is sampled. Each sample is converted into a digital word by forming a binary row of digits from it. The binary rows are then entered into the memory. In this process a high sampling frequency is of great importance, because it is necessary to satisfy the criterion of the first sampling theorem. If the frequency of sampling is less than $2f_{max}$, a hidden distortion occurs. To ensure that every criterion is met, the signal must be passed through a low-pass filter whose cut-off frequency is f_{max}. This filter is the one that indicates latent distortions (or the lack of them). Since even this filter cannot be ideal, to minimise the effects of a real filter the sampling frequency is raised to $3f_{max}$.

In this manner, the words which get into a memory of word capacity N at a frequency $3f_{max}$ will mean that the content of the memory represents a time period t, that is

$$t = \frac{N}{3f_{max}}. \tag{8.51}$$

It has been proved that to produce a statistically independent spectrum with two degrees of freedom, the analyser must sample at intervals of $1/\beta$. Thus the time period of the memory should also be $1/\beta$ i.e.

$$\frac{1}{\beta} = \frac{N}{3f_{max}}. \tag{8.52}$$

This will determine how many words the memory should be able to store, because β is equal to the resolution. In this way if the time-compression analyser produces a spectrum which includes M spectral lines, corresponding to a filter of constant bandwidth with M adjacent bands, then

$$\beta = \frac{f_{max}}{M}. \tag{8.53}$$

Comparison of (8.52) and (8.53) gives

$$M = \frac{N}{3}. \tag{8.54}$$

For instance, in the Brüel and Kjaer 3348 type narrow-band real-time analyser the adjacent filters – i.e. the spectral lines – number $M=400$. This gives a memory capacity of $N=1200$.

The signal to be measured is fed through the amplifier 1 to the filter set 2. The latter consists of M filters having mid-frequencies $f_1, f_2, f_3 \ldots f_M$, characteristics (in the frequency range of interest) close to each others, but pass-bands which do not overlap. With the selector switch the desired filter of the set can be connected to the indicator 4; or to the level recorder 5. The indicator and recorder make it possible, for example, to trace visually or record etc. the frequency components. Manufacturers usually design their real-time analysers in such a way that each channel (i.e. each frequency band) has its own effective value forming device 6 and its own memory 7. For example, the Brüel and Kjaer real-time octave analyser can obtain the following values from the signal:

(a) the r.m.s. value at the instant of interrogation of the channel

(b) the momentary r.m.s. value at the beginning of the interrogation cycle, at the same instant for each channel,

(c) the maximum r.m.s. value between two interrogations.

The switch marked 8 in the simplified block diagram of Fig. 8.23 allows one to send the interrogated values to the monitor 9 (which usually includes a large screen using electromagnetism to deflect the electronic beam), to the level recorder 11 and to the A/D transducer 12.

The time base generator 10 usually qaurtz-based causes the horizontal deflection on the screen of the monitor 9 to change cyclically, in synchronism with the switch 8 (dotted line, common control). Thus on the screen the momentary r.m.s values of the

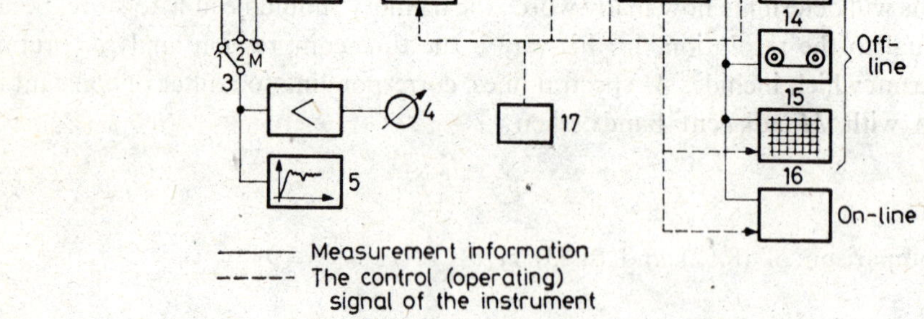

Fig. 8.23. *Simplified block diagram of a real-time analyser*

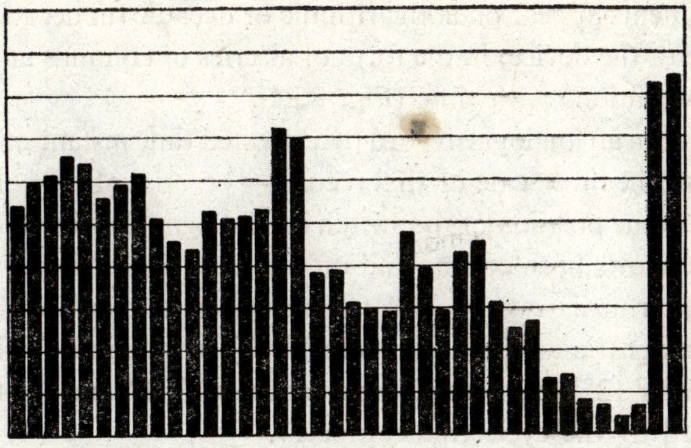

Fig. 8.24. *The screen of the monitor of a real-time analyser*

TABLE 8.3.

The main details of the Brüel and Kjaer 3347, 3348 and 2131-type real-time analysers

Type	3347-type octave real-time analyser	3348-type narrow-band real-time analyser	2131-type digital real-time frequency analyser
Frequency range	25 Hz...20 kHz 12.5 Hz...40 kHz (according to order)	0...10 Hz, 0...20 Hz etc. 0...20 kHz/to 1—2—5 stages	1.6 Hz...20 kHz third and 2 Hz...16 kHz octave
Number of band filters	31 third, 38 third (according to request)	400	42 third 14 + linear octave
Graduation of the frequency scale	logarithmic	linear	logarithmic
Dynamic range	64 dB	60 dB (averaging)	69 dB
Range of sceen	50 dB	50 dB	60 dB
Graduation of the amplitude scale of the screen	linear and logarithmic	linear and logarithmic	logarithmic
Comparison of spectrum	not possible	possible	possible
Signal input	analogue	analogue	analogue and digital
Analogue output	level recorder, $X-Y$ recorder (according to request)	level recorder, $X-Y$ recorder (according to request)	level recorder, $X-Y$ recorder
Digital output	given	given	given

frequency components appear on a logarithmic or decadic (in decades) scale (depending on the design of the device) in the form of a series of columns and allow the elucidation of their variations with time (Fig. 8.24).

From the spectrum an analogue record of a selected time instant can be obtained by the level recorder. The time scale of that record — because of the common control — is closely related to the position of the switch 8. The signal leaving the A/D transducer 12 the apparatus allows both on-line and off-line data processing. For off-line operation, however, intermediate storage is necessary, on a punched tape via the tape punch 13, on a magnetic tape via the recorder 15, or else in a ferrite-core memory. The A/D transducer 12, the intermediate memories 13, 14 and 15, the computer 16, etc. are controlled by the cycle transmitter 17.

Cyclic interrogation at high frequency saves considerable time.

Real-time analysis is not only the only way to examine transient signals (e.g. in shock analysis), but it also offers considerable advantages in the examination of stationary (deterministic or stochastic) signals and processes. Other examination methods may also be used, but the advantages offered by real-time analysis, namely very short examination times and more detailed and more exact information about the frequency composition, make it superior to all other known methods.

As can be seen, the filter sets of real-time analysers, depending on the bandwidth, may include octave, third or narrow-band filters. The bandwidths of the filters determine the resolution. The analysers on the market use third or narrow-band filters, and less frequently octave filters.

By way of example, Table 8.3 shows the most important details of three Brüel and Kjaer analysers, all well known and widely used.

9. EVALUATION OF STOCHASTIC MEASUREMENT INFORMATION

Chapters 3 and 4 dealt with the properties of non-deterministic measurement information — non-deterministic signals — and some of the principles involved. In this chapter, using the principles previously treated and with special attention to the requirements of practical measurements, we shall discuss the more important methods of evaluation.

9.1 BASIC CONCEPTS, GENERAL PROBLEMS

Numerous phenomena in everyday life, technology, scientific research and research into economics, necessitate stochastic methods for their control and their evaluation.

A complete enumeration would go beyond the scope of this book, so we shall outline a few of the applications of stochastic methods.

Stochastic methods are required in the following fields:

Mechanics:

in strength analysis:

— in the stress calculation and testing of structures under dynamic loads,
— in the theoretical examination of, and in experiments and tests on, the composition of loads, and load accumulation,
— in the analysis of stresses caused by composite loads (e.g. internal combustion engines, electric motors, prime movers, etc.),
— in laboratory and plant test on deformation and flexibility,
— in the examination of fatigue phenomena in structural materials (metals, plastics, construction materials, etc.) .

in vibrations and oscillations:

— in the examination of vehicle suspensions and their behaviour on bumpy roads,
— in the examination of the safety of travel — extending also to the safety of vehicle suspensions,

— in the examination of the biological effects of vibrations and oscillations on humans. In addition to the great technical and physical importance of these problems, their biological and medical implications are also of major significance. It deserves note that the causal relations of the biological effects on humans occur because of their acceleration (more precisely the r.m.s. value of acceleration) and not because of the amplitude of the oscillations or vibrations.

in fluid mechanics:

— in the measurement of stochastically varying pressures,
— in the examination of the complex phenomenon of turbulence and its related characteristics,

in acoustics:

— in the measurement of problems related to noise from the aspect both of the source and of abatement,

in information transfer and telecommunications:

— in the handling, converting, transfer, etc. of information via telecommunication channels contaminated by noise,

in biology, environmental protection etc.:

— in the analysis of biopotentials,
— in the examination of the concentration of air pollutants.

Even from this far from complete list the reader will see that statistical methods are used in a large number of widely differing fields.

It may be stated that in general these methods — be they calculations or experimental measurement processes — replace the earlier empirical computing or measuring methods by reliable prognoses based on statistics, and on concrete data obtained by objective and reliable tests.

In what follows, we shall go into the concrete physical phenomena, because the statistical measuring devices and methods operate exclusively with physical quantities characterised by electrical signals. We shall not now deal with the conversion, assuming that conversion has already taken place. First some concepts will be revised and then we shall introduce some new notations and concepts.

In Section 4.2 we treated some general problems of stochastic processes, with some stress on stationary stochastic processes. A stochastic process was defined as a single-parametric mass of certain probability variables, and it was stated that in the cases examined by us the role of the parameter would be taken by time.

Although mathematically precise, this definition does not illustrate the practical measurement of stochastic signals with sufficient clarity. To prevent difficulties in understanding, we will provide meanings for "stochastic measurement data", and "stochastic measurement signals", which will serve in practice. It should, however, be borne in mind that our formulation does not replace an exact mathematical definition, it only complements it.

The most important characteristic of a stochastic signal carrying stochastic measurement information, which distinguishes it from deterministic signals, is the fact that the value of the function (the informative parameter) cannot be predicted for any value(s) of the independent variable(s) (independent informative parameter(s)).

It should be stated at this point that, although the statements about dimensions apply to stochastic signals just as much as they do to deterministic ones (Chapter 3) — that is, up to four-dimensional stochastic measurement information can occur. We shall concentrate on stochastic functions (stochastic signals) having a single independent variable, that is, on bi-dimensional functions and signals, because those with a higher number of dimensions have little practical significance.*

Most stochastic signals which occur in measurement practice, as already stated, are functions of time. The signals to be examined are similar, with the exception that our results can also be generalised to other time-independent parameter(s).

Stochastic time functions — hereinafter simply called "functions" — are known by other terms not only in the literature but also in measuring practice (e.g. random functions, statistically varying functions, stochastic functions, etc.).

Stochastic functions can be described and written down in terms of probability data alone. The probability data, also called statistical characteristics, — mean value, distribution function, distribution density function, correlation function, standard deviation, etc. — can be determined by statistical analysis.

Statistical analysis is a complex process used in measurement and measurement evaluation, in the course of which the statistical characteristics of the stochastic function are obtained from measurements already recorded or currently being made (in real-time analysis). In most practical cases the stochastic process is stationary. This, as shown in Section 4.2 facilitates the process. Since the ergodic hypothesis is also applicable in general, the stochastic function can be substituted by any one of its easy-to-determine recording.

Less frequently, non-stationary stochastic functions may also occur in practical measurements. Since their precise mathematical description is much more difficult, we shall not deal with their theory but give a brief outline of their practical examination.

We shall first study the statistical analysis of stationary stochastic functions (signals).

* We wish to note that one-dimensional measurement information (which carries random fluctuations of information in measurements) has already been treated in Chapter 6. In that treatment we applied probability and mathematical-statistical methods, as we shall also do here.

9.2 STATISTICAL ANALYSIS

In Section 4.2, the reader became familiar with the stochastic function and the distribution density functions of the stationary stochastic process, on the basis of which the statistical characteristics of mean value, correlation function etc. have been interpreted.

Let us therefore set out from the assumptions that only records of measurement are available for examination, that the statistical characteristics cannot be derived from the distribution or the distribution density functions because these are not available, and that we can only draw conclusions about the distribution from the records.

9.2.1 EXAMINATION OF RECORDS

Let us assume that the process examined is stationary and ergodic. This means that the arbitrary records $x^{(1)}(t)$, $x^{(2)}(t) \ldots x^{(i)}(t)$ are statistically equivalent to one another and with the stochastic function $x(t)$ as well.

For practical measurements it is not advisable to take a large number of records because this would involve too much information which would be difficult to handle. It is usually sufficient to study one or two samples, provided that they cover a sufficiently long period of time and that the statistical characteristics derived from them are in agreement with the statistical characteristics obtainable by data reduction (cf. Section 6.2.2).

The stationarity and ergodicity of the process – which latter is generally accepted without further verification – permit the essential simplifications mentioned before. To make a rough check on the ergodicity of the process, the ensemble-average and time average of a function recorded over a sufficiently long period of time are examined and compared. If the two are in fair agreement, depending on the accuracy required, the process may be regarded as ergodic.

In Section 4.2 we presented the mathematical definition of the probability distribution, together with the corresponding distribution and density functions. This definition includes an illustrative geometric interpretation related to, and interpreted on the basis of, the record. We remind the reader that the single-variant i.e. first probability distribution that is density function is given in (4.33) and (4.34) while the bi-variant, i.e. second-distribution and density function, is given in (4.35) and (4.36).

For a single record, the single-variant distribution functions give the probability that the value of the $x(t)$ under study, that is, the arbitrary ith $x^{(i)}(t)$ sample which is statistically equivalent to it at time t_1 is less than x.

The bi-variant distribution function gives the probability that at time t_1 $x(t)$ is less than x_1 and at time t_2 it is less than x_2.

The density functions can be interpreted as the derivatives of the corresponding distribution functions (cf. 4.34 and 4.36). This means that e.g. from the single-variable distribution function $F_{t1}(x)$ the single-variable density function $f_{t1}(x)$ can be obtained by producing the boundary value of the well known change ratio of

$$f_{t1}(x) = \lim_{\Delta x \to 0} \frac{F_{t1}(x) - F_{t1}(x+\Delta x)}{\Delta x}. \tag{9.1}$$

Similar considerations may relate to the bi-variant case, but since for the time being we need only the single-variable distribution and density functions, we concentrate on these.

Sufficiently long time duration

As has been seen, the stationary stochastic process can be described by a single sample recorded over an infinitely long time period. Since, however, this is impossible in practice, we must be satisfied with records for a sufficiently long time period. There is no accurate mathematical term for the sufficiently long time period, since this concept is an approximation 'ab initio'. However, on the basis of the conditions of measurement and of practical considerations in some cases its value can be determined with a suitable precision. There is no general formula for its determination: it depends on the local conditions, on the results of tentative measurements or records, and on a certain amount of practice which will enable anybody to determine the duration of recording which is sufficient to ensure the desired accuracy. Let us denote it by T.

Figure 9.1 shows the method of interpreting and deriving the probability distribution and density functions of the process on the basis of the realisation of the stochastic process recorded for the duration T.

The distribution function is interpreted on the basis of Fig. 9.1 (a). According to the definition (4.33), the distribution function gives the probability that at the instant(s) of examination the record will have a value less than a given value of x. In Fig. 9.1,

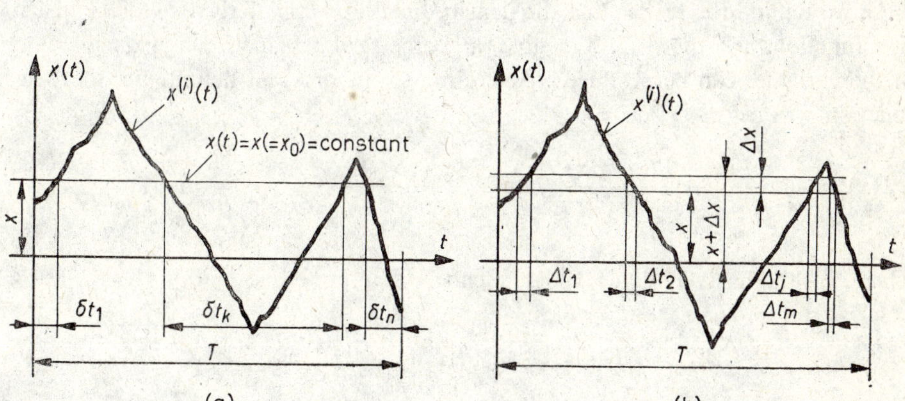

Fig. 9.1. *Derivation of the probability distribution and density functions from a single sample of the stochastic process recorded during a time period T:* (a) *distribution function,* (b) *interpretation of the density function*

we drew for that value of x, parallel to the t-axis, a straight line, whose equation is:

$$x(t) = x(= x_0) = \text{constant}. \tag{9.2}$$

It is evident that the complexity of the time instants — i.e. the total time period — within the examination time T which are "favourable" for probability, i.e. where the record is less than x, is equal to the sum of the indicated values δt_k, i.e.

$$\delta T_n = \delta t_1 + \delta t_2 + \ldots + \delta t_k + \ldots + \delta t_n = \sum_{k=1}^{n} \delta t_k. \tag{9.3}$$

In agreement with the definition (4.33) and the definition (4.1) of probability, the distribution function in this case is given by the ratio of the sum of the favourable time periods to the maximum possible duration, i.e. the overall examination time T. Thus

$$F(x) = P[x(t) < x] = \frac{\delta T_n}{T}. \tag{9.4}$$

Note that since the distribution function does not refer to a single time instant, the previously applied index has been omitted and we shall omit the density function as well.

For the sake of completeness, we should note here that the complementary function of the distribution function, denoted by $F_{\text{compl}}(x)$, can also be interpreted in the following way:

$$F_{\text{compl}}(x) = 1 - F(x). \tag{9.5}$$

In certain cases this interpretation could be essential, because some instruments are based on the determination of the complementary distribution function.

We shall now define the density function, referring to Fig. 9.1b. With reference to (9.1), we remind the reader that the density function is the derivative of the distribution function, and that Fig. 9.1b gives a geometrical explanation how the increments of the variables can be produced. According to the previous definitions, the density function will be

$$f(x) = \lim_{\Delta x \to 0} \frac{F(x) - F(x + \Delta x)}{\Delta x} = \lim \frac{P[x < x(t) < x + \Delta x]}{\Delta x} = \lim_{\Delta x \to 0} \lim_{T \to \infty} \frac{1}{T} \frac{\Delta T_m}{\Delta x}, \tag{9.6}$$

where, on the basis of Fig. 9.1b, ΔT_m can be interpreted as

$$\Delta T_m = \Delta T_1 + \Delta T_2 + \ldots + \Delta T_j + \ldots + \Delta T_m = \sum_{j=1}^{m} \Delta T_j. \tag{9.7}$$

Both (9.4) and (9.6) are precise definitions. As regards (9.6), we have to add that because of the two limits we must accept a compromise as to accuracy in practical measurements. Instead of recording over an infinitely long time, we must be satisfied with recording over a "sufficiently" long time period: the so-called "window-width" Δx is also finite, i.e. small. In instruments it amounts to approximately one tenth of the r.m.s. value of the stochastic record, sufficient to ensure a quite accurate measurement and evaluation.

Classification, classifying methods

The values of stochastically varying signals are determined at certain time instants t_i and classified into value groups. This in fact is the essence of classification. It enables us to gain useful information about the properties of the stochastic signal.

The signal values can be determined by interrogation of the record, i.e. by measuring the signal value at the time instant concerned.

Classification may be

— analogue,
— digital.

In analogue classification the values of the functions are measured by an analogue device which interrogates the record at defined time instants. The measured analogue value determines the group to which a given function point belongs.

In digital classification, the function values corresponding to the time instants examined are quantized, i.e. they fall between the two boundary values of quantization, and the function value corresponding to the arbitrary time instant t_i falls into one of the classes.

Figure 9.2 illustrates the principle of the classification of a given recording. In classification, the recording of the function is interrogated at suitable instants of time. Interrogations may take place at regularly spaced instants t_i (Δt apart). In Fig. 9.2, the classes are separated by values of x which correspond to lines parallel to the t-

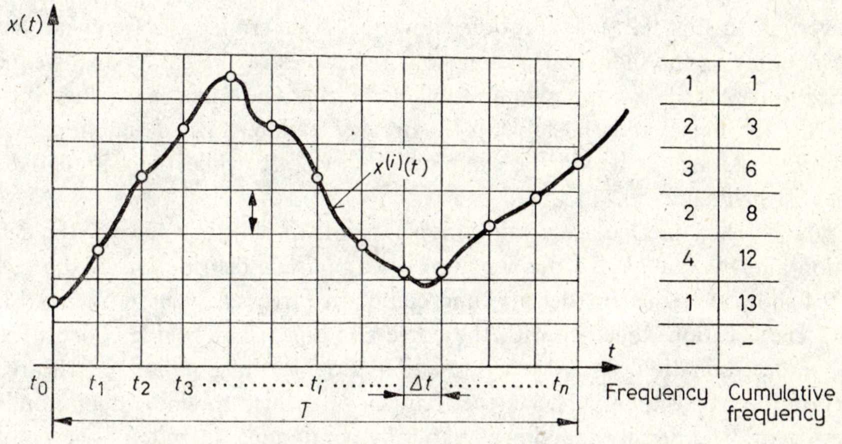

Fig. 9.2. *The principle of the classification of recordings*

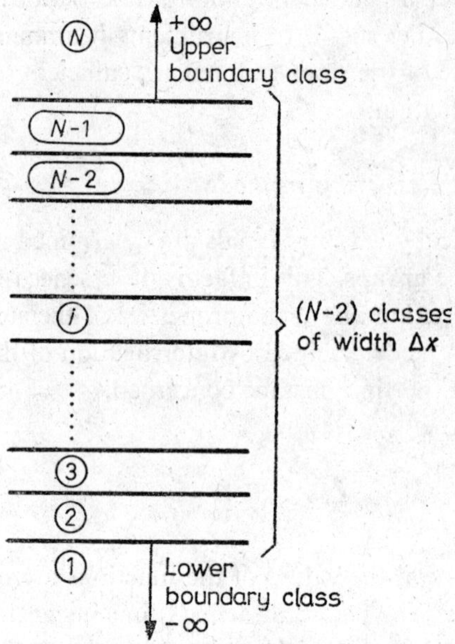

Fig. 9.3. *Classes in a classifying process*

axis, and differ from adjacent values by Δx. The number of classes depends on the nature of the classifying apparatus, and on its specifications, e.g. the width of the classes, the dimension of the measurand physical quantity and the number of classes. For example, in a device with $N=12$ classes there are $N-2=10$ classes of width Δx plus a lower and an upper so-called boundary class, whose extent in one direction is infinite. The boundary classes are naturally contiguous with the adjacent classes (see Fig. 9.3).

Reverting to Fig. 9.2: classification allows us to determine the frequency values corresponding to the individual groups and the values of the cumulative frequency (cf. Section 6.2.2). They are shown in tabulated form. The frequency value, as is evident from the figure and the table, is the number of interrogated functional values falling into the groups within the examined interval T, while the cumulative frequency is their sum.

On the basis of the frequency values, it is possible to plot the values of the density function and the cumulative frequency of the amplitude distribution of the record. Fig. 9.4 shows a frequency (density) and cumulative frequency curve obtained from such a classification, together with the corresponding record which constitutes the basis of determination. The density function provides information on the record $x^{(i)}(t)$ on the left of the figure, i.e. the frequency at which the individual amplitude values occur. The density function is naturally continuous, in spite of the fact that during classification only its discrete points are plotted.

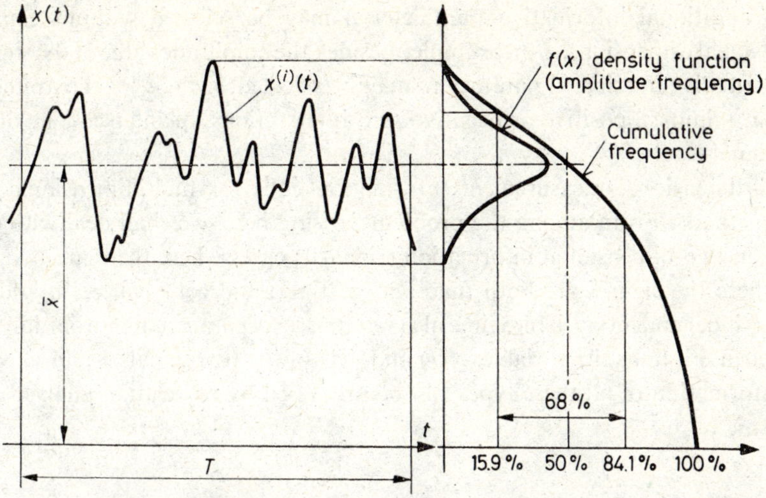

Fig. 9.4. *Recording of the measured signal with the density function of the corresponding amplitude distribution and a cumulative frequency curve*

The greater the number of classes N and the higher the frequency of interrogation of the functional values, i.e. the smaller Δt is, the greater the resolution of the discrete values. The width of the bell-shaped curve which corresponds to the density function $f(x)$ is equal to the peak-to-peak value of the record examined (the difference between the greatest and smallest values occurring in the interval examined).

The type of distribution may vary but in practice stochastic processes with normal (Gaussian) amplitude distributions are especially common. In these cases, as was assumed in relation to Fig. 9.4, the arithmetic mean corresponds to the 50% abscissa value of the cumulative frequency curve, while the r.m.s. of the recording $x^{(i)}(t)$ (mean square value) corresponds to the 15.9% abscissa value.

Classification parameters, dimensional problems

The classification process illustrated in Figs. 9.2 and 9.4 was bi-dimensional and single-parametric. The dimension of the process is in agreement with the idea of signal dimension (the dimension of measurement information) introduced in Chapter 3, since it carries bi-dimensional information: $x^{(i)}(t)$. Therefore, the classification process is also bi-dimensional.

The parameter of classification, as a notion, is complementary to the information parameter(s) because e.g. in Eq. (3.1a) and Eq. (3.1b) the independent variables of the function x, y, z were called information parameters while here the function proper, i.e. x is regarded as the parameter. In general, in the classification of bi-dimensional measurement information, the classification is one-parametric. It might however happen that, in addition to the main informative parameter, one (less frequently several) additional informative parameter(s) may be selected without causing increased signal dimensions. For example, besides the amplitude value of the signal, further informative (auxiliary) parameters may be selected, such as local extreme values (maxima, minima), positive or negative zero intersections, or intersections of certain given signal-levels.

Multi-dimensional measurement information calls for multidimensional classification methods. From among the problems arising here, we shall deal with cases of more than two-dimensional information (signals); cases where there are two signals; cases where the signals are both functions of time; and cases where in addition to being an independent variable, time also features as a parameter (not an informative one), because it links the variables $x(t)$ and $y(t)$ of the two signals x and y. Measurement information of all these types can be processed by correlation analysis (cf. Section 9.3).

9.2.2 INSTRUMENTAL DETERMINATION OF THE AMPLITUDE DISTRIBUTION OF RECORDS

In the previous section the principles of evaluation of stochastic recordings were dealt with. The subject is very important, because it serves as a basis for those practical evaluation methods which measure from point-to-point, i.e. the old "small-scale" methods which are applied in automated form in modern measuring instruments.

Despite the common basic principles, there are actually several practical methods for the determination of the amplitude distribution and other statistical properties of stochastic signals and records.

One of the most efficient methods is real-time analysis (Section 8.4.4). However, at least with digital or hybrid processes, both the x and the t co-ordinates must be simultaneously and frequently scaled and, although devices for the purpose are commercially available, those with sufficient resolution are very expensive. Therefore their more widespread use can be expected only in the future.

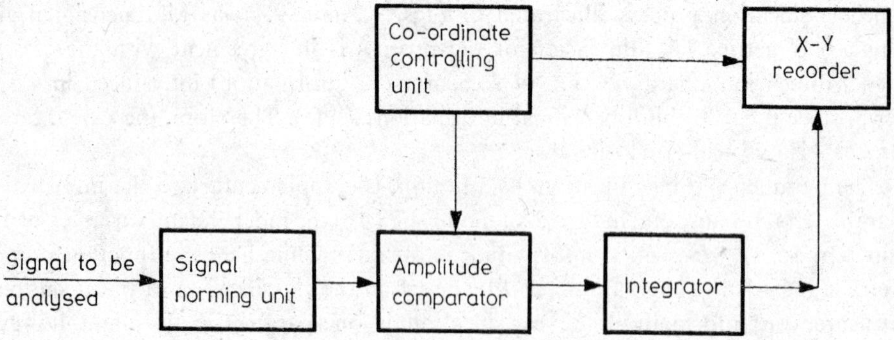

Fig. 9.5. *Block diagram of a widely used variant of the amplitude distribution analyser*

A more frequent solution is the use of devices which evaluate fully automated.

Figure 9.5 is the block diagram of a widely used amplitude distribution analyser system. Its structure is similar to that of the well known DISA 52B00 amplitude distribution analyser (the so-called APA, Amplitude Probability Analyser, System) [9.5] developed primarily for the analysis of stochastic records obtained from the measurement of turbulent flows. In its more advanced form the system is suitable for handling and processing all sorts of stochastic records.

Using the block diagram in Fig. 9.5 we shall study the operation of this system. The stochastic record to be analysed reaches the signal norming unit first. That is mounted on the input of the system. That unit simplifies it and converts the stochastically varying signals into uniform ones. "Unit signal" means uniform amplitude range, but uniformisation generally takes place by the r.m.s. value. For example, a unit signal of r.m.s. value 1V means that the r.m.s. value corresponding to the highest peak-to-peak value in the range of investigations is 1V, and the rest of the amplitude values are scaled correspondingly. The signal norming unit is fitted with an accurate – usually digital – volt meter, to measure the unit signal values corresponding to the input signal.

Amplitude comparator

The amplitude comparator is the most important unit that is, subsystem in the system which, in fact, performs the analysis. Into its input a previously normed signal is fed, while at its outputs various frequency- and time-period-modulated pulses appear, from which the required distribution and density functions of the amplitude distribution can be derived with ease.

The simplified functional block diagram of the amplitude comparator is shown in Fig. 9.6. Its two main inputs are the signal and the level inputs. The stochastic signal to be analysed is sent to the signal input. At the level input (marked "x_0") that constant voltage of value x_0 with reference to which the probability distribution and the density functions are to be defined (cf. Fig. 9.1) is set. Depending on the device, this value can be set at any desired magnitude within the normed amplitude range.

The window-width adjusting element, marked Δx, of the so-called "window-generator" may also be called an input because it can be used to adjust the selected window-width which is given as a fraction of the maximum value of the normed amplitude (e.g. 0.1 V; 0.15 V; 0.3 V etc.).

Knowledge of the shapes of the different signals in Fig. 9.7 will help in understanding the operation of the amplitude comparator. It shows that the amplitude comparator produces the threshold levels which can then be interpreted.

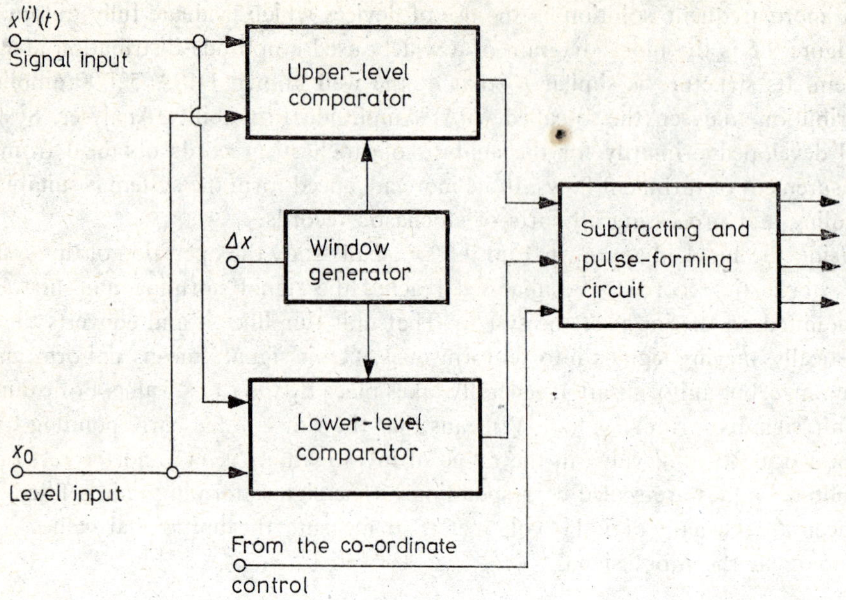

Fig. 9.6. *Simplified block diagram of the amplitude comparator of an amplitude distribution*

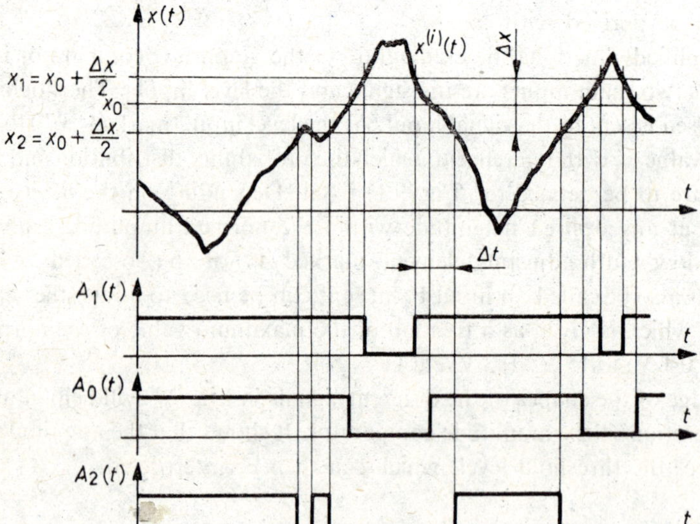

Fig. 9.7. *Signal shapes ('threshold levels') produced by the amplitude comparator*

Determination of the distribution function $F(x)$

As shown in Fig. 9.1a, to determine the distribution function $F(x)$ only $A_0(t)$ needs to be known. For this purpose it is sufficient to use one of the two (lower- and upper-level) comparators; one of them, say the lower, and leave the upper-level comparator out of operation.

The mean value of the function $A_0(t)$ is proportional to the distribution function $F(x_0)$ sought. Thus, after pulse forming, the function $A_0(t)$ is led to the integrator, the integrator's output signal, which is proportional to the mean value $A_0(t)$, together with the signal generated by the coordinate control unit, control the XY recorder. The recorder then plots $F(x)$ on their basis.

Determination of the density function $f(x)$

In agreement with Fig. 9.1b, the lower- and upper-level comparators generate a window of width $\Delta x/2+\Delta x/2=\Delta x$ and produce the functions $A_1(t)$ and $A_2(t)$ (threshold-level pulse series).

In the SINGLE mode of operation, the difference between the pulse series $A_1(t)$ and $A_2(t)$ is carried to the output. The mean value of the resultant pulse series (average) is proportional to $f(x)\Delta x$.

In the DUAL mode of operation, the mean value of the signals generated by both level comparators is calculated first, before the mean value of the pulse series is produced. This mode offers the great advantage of a very high upper boundary frequency.

Integrator

The integrator gives the mean value of the signals generated by the amplitude comparator. From the circuital point of view, it may be a conventional, i.e. RC, or a real-time integrator (cf. Section 8.4.4). The real-time integrator is much more expensive than the conventional type, but it integrates approximately four times faster and very accurately, and is therefore also suitable for handling non-stationary signals.

Co-ordinate control unit

The co-ordinate control unit generates the control signal in the Y direction of the $X-Y$ recorder. Jointly with the integrator which produces the signal in the X direction, it controls the recorder.

Summing up, the system shown in Fig. 9.5 is suitable for the evaluation of $x^{(i)}(t)$ stochastic signals and for the direct determination of their amplitude distribution, and their distribution and density functions. In addition to this system there are others for the purpose with different layouts, e.g. real-time devices. The differences between them are either insignificant or only structural and not functional. This means that, since every system sets out from the mathematical definition, they all include a window generator, a level adjusting facility, etc. Our goal was merely to outline a number of important principles. We chose this system because it features in several devices which have proved their value in practice.

Notes on the determination of the amplitude distribution of non-stationary stochastic signals

Our discussions of the principles of stochastic signals in Chapter 4 and in the present part of the book were based on the assumption that these signals are stationary. This assumption applies to various stochastic signals met in measuring practice and means that despite their constant and unpredictable variations, the average properties over a longer period of time — which we called a "sufficiently long time" — are constant.

The statistical properties even of non-stationary signals can be determined, provided that the integrator doing the averaging (e.g. the one in Fig. 9.5) takes the average for a short period, during which the average properties of the signal show only negligible changes. This is possible if e.g. the integrator works on the real-time principle (cf. Section 8.4.4). In such a case even non-stationary signals can be analysed. Ref. [9.6] describes such analysis.

9.3 CORRELATION ANALYSIS

In Section 4.2 we summarised the basic concepts of correlation, with stress on the mathematical problems of greater significance. Now its practical aspects and the devices used will be considered.

Correlation analysis is a modern method of evaluation. Although its principles and its mathematical basis are not new, its introduction into measurement practice has been an achievement of the last 20 to 25 years.

9.3.1 PRINCIPLES AND PROBLEMS OF THE CORRELATION TECHNIQUE

A correlation function can be defined in three ways: (i) it can be regarded as the mean value over time of a stationary process. (ii) it can be derived by means of probability calculus, and defined as the expected value of the product of two functional values taken at different instants of time; (iii) it can be spectrally interpreted.

In Section 4.2, although not explicitly, the first two definitions were referred to and compared with one another. It was stated — naturally for stationary processes only — that the ensemble average and the mean value over time are equivalent.

The third, equally important, definition was not mentioned in Section 4.2. It will now be treated briefly, without verification or derivations.

First of all, we wish to remind the reader of the Fourier and inverse Fourier transforms defined in Eqs. (3.15) and (3.16.) Let us subject the auto-correlation and cross-correlation functions defined in Section 4.2 to Fourier or inverse Fourier transformation. We get the transforms

$$\Phi_{xx}(j\omega) = \int_{-\infty}^{+\infty} \varphi_{xx}(\tau)e^{-j\omega t}\,\mathrm{d}\tau, \tag{9.8}$$

$$\Phi_{xy}(j\omega) = \int_{-\infty}^{+\infty} \varphi_{xy}(\tau)e^{-j\omega t}\,\mathrm{d}\tau, \tag{9.9}$$

and the inverse transforms

$$\varphi_{xx}(\tau) = \int_{-\infty}^{+\infty} \varphi_{xx}(j\omega)e^{j\omega t}\,\mathrm{d}\omega, \tag{9.10}$$

$$\varphi_{xy}(\tau) = \int_{-\infty}^{\infty} \varphi_{xy}(j\omega)e^{jwt}\,\mathrm{d}\omega. \tag{9.11}$$

Although stochastic signals do not have Fourier transforms, their correlation functions — not being stochastic functions with their variable τ — do have Fourier transforms, namely Φ_{xx} and Φ_{xy}, whose values are determined by the definitive equations (9.8) and (9.9). The transformed correlation functions Φ_{xx} and Φ_{xy} are also known as "power density spectra" (less frequently as "power spectra"), being related to the power of the signal $x(t)$.

The third possibility of generating correlation functions, not dealt with so far, is defined by the inverse transforms (9.10) and (9.11). From the power density spectra Φ_{xx} or Φ_{xy}, the correlation functions φ_{xx} or φ_{xy} can be determined. Note that the definitions (9.8) and (9.11) are known in the literature as the Wiener–Chintchin relations., see Refs [9], [11], [12], [9.2], etc.

The mean square of the stochastic signal $x(t)$ and the $x^{(i)}(t)$ record or sample can also be expressed by the power density spectrum as

$$\overline{x^2(t)} = \overline{x^{(i)2}(t)} = \frac{1}{2\pi} \int_{-\infty}^{\infty} \Phi_{xx}(j\omega)\,\mathrm{d}\omega. \tag{9.12}$$

The mean value of the product of the two stochastic signals x and y is

$$\overline{x(t)y(t)} = \overline{x^{(i)}(t)y^{(k)}(t)} = \frac{1}{2\pi} \int_{-\infty}^{\infty} \Phi_{xy}(j\omega)\,\mathrm{d}\omega. \tag{9.13}$$

9.3.2 THE MEANS AND THE EXECUTION OF CORRELATION ANALYSIS

The reader may have seen from the foregoing that, to create the means for some complicated analysis (frequency, real-time, amplitude distribution analysis etc.), experts always set out from a mathematical model, however sophisticated it may be, and try to implement it with the technology available. This also applies to the development of the means of correlation analysis. The real upswing in the field has come only in the last 15 to 20 years. Although a multitude of devices have been designed and built, and many of them are commercially available in extensive use, development goes on and new, better, cheaper instruments keep appearing.

As regards the mode of operation and the structural principle, correlation devices ("correlators", sometimes called "correlatographs") may be

— analogue
— digital or
— hybrid.

Here is a brief overview of the possibilities of producing correlation functions. Let us first examine the way in which the functional problems that follow from the mathematical definitions of the correlation functions influence the structural layout of a correlator, whether an analogue digital or hybrid device is concerned, Fig. 9.8, based on the definitions (4.51) and (4.52) of the auto- and cross-correlation functions, shows all those units which generate the functions. The reader will note that the correlator has many features similar to those of the amplitude distribution analyser illustrated in Fig. 9.5.

The signal is usually stored in a special magnetic tape unit, which is suitable for both signal storage and time delay.

The time delay unit is usually an electromagnetic propagation delay chain. At lower frequencies, special measuring tape recorders are used. Because of their higher precision these devices are costly, but they ensure signal storage and time delay (e.g. fast pickup and slow replay, or vice versa).

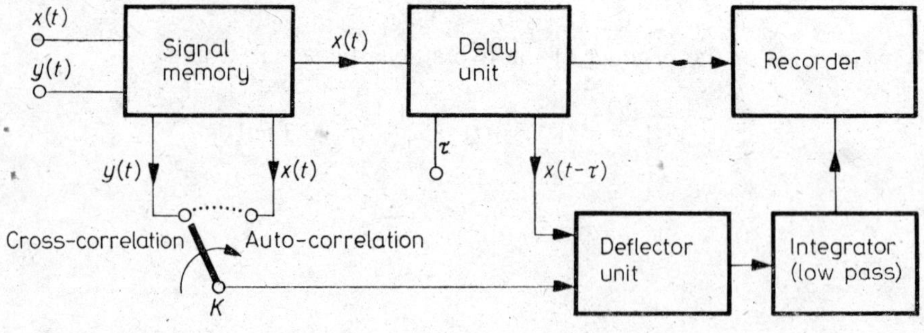

Fig. 9.8. *A correlator made up of functional units: K — operation mode switch, x(t) y(t) — input signals*

The multiplier unit may include a multiplier element based on some physical effect (e.g. a Hall generator) or a microelectronic multiplier circuit, etc.

With reference to the schematic diagram of the correlator (Fig. 9.8), to record correlation functions, the value of τ on the time delay unit must be varied within the measurand range, in order to obtain the total set of values of $\varphi_{xx}(\tau)$ and $\varphi_{xy}(\tau)$. This is done by a so-called sweep generator which scans the complete range continuously. A device with such a design is called a series or sequential correlator.

The time delay τ can also be varied by using a large number of fixed-value time-delaying members in series, which in fact cover the entire range of τ. The layout of these devices is similar to that of the real-time devices, because they produce the co-ordinates of the correlation function which are represented by different values of τ, simultaneously. Such devices are called parallel correlators.

Analogue correlators

Analogue correlators are usually serially operating. They are suitable for the examination of low-, medium-, and high-frequency signals. The low-frequency analogue correlators work almost exclusively with magnetic-type time lag, while the high-frequency variants apply time-lags generated by TC elements.

Digital correlators

Digital correlators work on sample signals obtained through the interrogation of stored records, at a certain frequency. In this case the cross-correlation function is produced by the following formula:

$$\varphi_{xy}(\tau) = \frac{1}{N} \sum_{k=0}^{N} x(k\Delta t - \tau)\, y(k\Delta t). \tag{9.14}$$

To achieve the required resolution, frequent sampling is necessary. The electrical circuits of digital correlators are therefore rather complicated.

Considerable efforts have been made in recent years to accelerate and automate the data-processing that occurs in the determination of correlation functions. An outstanding result of these efforts has been the FFT (Fast Fourier Transform) method. In carrying out the iterations involved, valuable assistance is lent by the latest computer applications, e.g. microprocessors, which are now available at moderate prices.

Hybrid correlators

With the increasing frequency of sampling, the number of electronic elements, and higher accuracy of devices, their price is rising rapidly (non-linearly). Hybrid devices are capable of only a rough resolution of the records and usually operate in parallel. Their price and accuracy lie somewhere between those of their analogue and digital counterparts.

Summing up, correlation methods assist in the solution of problems where one or several randomly changing (stochastic) physical quantities, their relationships or their individual or collective properties are to be examined. Such problems occur in aero- and hydro-dynamics, the examination of oscillations, in acoustics, geophysics, radio astronomy, biology, etc.

Correlation analysis calls for sophisticated electronic methods which, owing to the rapid progress made in recent years, are now available.

Readers interested in correlation methods can find abundant information in the literature, e.g. in Refs [9.2], [9.3], [9.4], [9.7], [9.9].

REFERENCES

(A) GENERAL AND RECOMMENDED BOOKS WITH REFERENCES TO SEVERAL CHAPTERS OF THE BOOK

[1] Ambróczy, A. and Jávor, A.: *Evaluation of Measurement Data*. (In Hungarian). Programming-Electronics Series. Műszaki Könyvkiadó, Budapest 1976. p. 157.

[2] Boros, A.: *Electrical Measurements in Engineering*. Studies in Electrical and Electronic Engineering. No. 17. Elsevier, Amsterdam 1985.

[3] Bronstejn, I. N. and Semendjaew, K. A.: *Mathematical Pocketbook*. (In Hungarian). Műszaki Könyvkiadó, Budapest 1982.

[4] Broch, J. T.: *Application of B and K Equipment to Mechanical Vibration and Shock Measurements*. Brüel and Kjaer, Naerum 1973. p. 311.

[5] Broach, J. T.: *Application of B and K Equipment to Acoustic Noise Meaurements*. Brüel and Kjaer, Naerum 1971. p. 203.

[6] Brüel and Kjaer: *Master Catalogue*. Naerum 1974.

[7] Brüel and Kjaer: *Application of Brüel and Kjaer Equipment to Power Spectral Density Measurements*. Naerum 1972. p. 57.

[8] Brüel and Kjaer: *Vibration Course*. (In Hungarian). Naerum 1975.

[9] Csáki, F. and Bars, R.: *Automatics*. (In Hungarian) 4th Ed. Tankönyvkiadó, Budapest 1983. p. 702.

[10] DISA (Dansk Industry Syndicat AS) Skovlunde: Informational material on the products. (The name of the company has been changed to DANTEC Elektronik as from July 1, 1983).

[11] Ferenczy, P.: *Communication Theory*. (In Hungarian). Tankönyvkiadó, Budapest 1974. p. 251.

[12] Fodor, Gy.: *Analysis of Linear Systems*. (In Hungarian). Műszaki Könyvkiadó, Budapest 1976. p. 312.

[13] Frank, H.: *Electrical Measurement Analysis*. McGraw-Hill, New York 1959.

[14] Fritzsche, G.: *Informationsübertragung*. Verlag Technik, Berlin 1977.

[15] Hofmann, D.: *Handbuch. Meßtechnik und Qualitätssicherung*. 2nd ed. Verlag Technik, Berlin 1981. p. 471.

[16] Hofmann, D., Meinhard, R. and Reineck, H.: *Meßwesen, Prüftechnik, Qualitätssicherung. Begriffe und Definitionen*. Verlag Technik, Berlin 1980. p. 323.

[17] Holzweissig, F. and Meltzer, G.: *Meßtechnik der Maschinendynamik*. Fachbuchverlag, Leipzig 1973. p. 412.

[18] Jakowlew, K. P.: *Mathematische Auswertung von Meßergebnissen*. Verlag Technik, Berlin 1952.

[19] Jánossy, L.: *Theory and practice of the evaluation of measurement results*. (In Hungarian). Akadémiai Kiadó, Budapest 1968. p. 527.

[20] Korn, G. A. and Korn, T. M.: *Mathematical Handbook for Scientists and Engineers*. (In Hungarian). Műszaki Könyvkiadó, Budapest 1975. p. 995.

[21] Richter, W. (ed.): *Fachlexikon Meßtechnik*. Fachbuchverlag, Leipzig 1984. p. 528.

[22] Tarnay, K.: *Electronic Measurements*. (In Hungarian). Műszaki Könyvkiadó, Budapest 1962. p. 466.

[23] Woschni, E. G.: *Meßdynamik*. S. Hirzel, Leipzig 1972. p. 194.

[24] Warnecke, H. J. and Dutschke, W.: *Fertigungsmeßtechnik*. Springer, Berlin 1984. p. 812.

(B) SPECIAL LITERATURE, WITH REFERENCE TO PARTICULAR CHAPTERS OF THE BOOK

CHAPTER 1

[1.1] The decree No. 8/1976 (IV. 27) of the Council of Ministers on Metrology. *Magyar Közlöny*, No. 34.

[1.2] MSZ (Hungarian Standard) No. 4800/1–11 The names, symbols and units of physical quantities (Standards).

[1.3] Meissner, F.: SI-Umrechnungstabellen. 2nd Ed. Fachbuchverlag, Leipzig 1979. p. 72.

[1.4] Cs. Pintér, P.: *Quantities, units, numbers – SI*. (In Hungarian). Műszaki Könyvkiadó, Budapest 1987. p. 700.

CHAPTER 2

[2.1] MSZ 18200—73 *Standard Process Instrumentation Terminology*. (In Hungarian). ISA–551. 1/1976.

[2.2] Mesch, F.: *Systemtheorie in der Meßtechnik — eine Übersicht*. *Technisches Messen ATM*. 1976/4. VOO—10. pp. 105–112.

CHAPTER 3

[3.1] Novickij, P. V. *et al.*: *Gütekriterien für Meßeinrichtungen*. Verlag Technik, Berlin 1978. p. 152.

[3.2] *Brockhaus ABC Automatisierung* (Lexikon). Brockhaus Verlag, Leipzig 1975. p. 739.

[3.3] Dobrinin, E. M. and Pan Čzun Čzen: Problems of Dynamical Accuracy of the Devices Used to Measure Length Automatically. (In Russian). Mashgiz, Moscow 1963.

[3.4] Lenk, A.: *Elektromechanische Systeme*. Vol. 3.: Systems with Auxiliary Power. Verlag Technik, Berlin 1975.

[3.5] Gutnikov, V. S., Lenk, A. and Mende, U.: *Sensorelektronik*. Verlag Technik, Berlin 1984. p. 260.

[3.6] Ornatskii, P. P.: *Automatic Measuring Instruments*. (In Hungarian). 2nd Ed. Műszaki Könyvkiadó, Budapest 1974. p. 531.

[3.7] Schnell, L. (Editor): *Technology of Measuring Signals and Systems*. (In Hungarian). Műszaki Könyvkiadó, Budapest 1985. p. 1112.

CHAPTER 4

[4.1.] Prékopa, A.: *Probability Calculus with Technical Applications*. (In Hungarian). Műszaki Könyvkiadó, Budapest 1980. p. 440.

[4.2] Fazekas, F. (Editor): *Technical Mathematical Exercises*. (In Hungarian). C/V Vol.: Medgyessy,

P. and Takács, L.: *Probability Calculus.* (In Hungarian). Tankönyvkiadó, Budapest 1957. p. 333.

[4.3] Heinhold, J. and Gaede, K-W.: *Ingenieur Statistik.* R. Oldenbourg, München–Wien 1968.

[4.4] Papoulis, A.: *Probability, random variables and stochastic processes.* McGraw-Hill, New York 1965.

CHAPTER 5

[5.1] Doebelin, E. O.: *Measurement Systems.* McGraw-Hill, New York 1966.

[5.2] Lenk, A.: *Elektromechanische Systeme.* Vols I–III. Verlag Technik, Berlin 1971—1977.

[5.3] Petrik, O.: *Systems Engineering.* (In Hungarian). Tankönyvkiadó, Budapest 1977. p. 205.

[5.4] Szücs, E.: *Similarity and Model.* (In Hungarian). Műszaki Könyvkiadó, Budapest 1972. p. 299.

[5.5] Csákány, A. and Vajda, F.: *Microcomputers.* (In Hungarian). 2nd Ed. Műszaki Könyvkiadó, Budapest 1977. p. 292.

[5.6] Kreul, H.: *Was kann mein elektronischer Taschenrechner?* (2nd Ed.) Fachbuchverlag, Leipzig 1979. p. 131.

[5.7] Gombkötő, P. and Kaboldy, P.: *Machine Design. Pocket Calculator Programs.* (In Hungarian). Műszaki Könyvkiadó, Budapest 1984. p. 154.

[5.8] Kreul, H.: *Programmierbare Taschenrechner.* Fachbuchverlag, Leipzig 1980. p. 198.

[5.9] Csákány, A.: *What can a pocket calculator do?* (In Hungarian). Műszaki Könyvkiadó, Budapest 1981. p. 160.

[5.10] Baky, M.: *Calculator Programs PTK 1050.* (In Hungarian). Műszaki Könyvkiadó, Budapest 1985. p. 207.

CHAPTER 6

[6.1] Leinweber, P. (Editor): *Pocketbook of Linear Measurements.* (In Hungarian). Műszaki Könyvkiadó, Budapest 1960. p. 952.

[6.2] Szombathy, E.: *Technical Measurements.* (In Hungarian). Felsőoktatási Jegyzetellátó V., Budapest 1957. p. 284.

[6.3] Warnecke, H. J. and Dutschke, W.: *Fertigungsmeßtechnik.* Handbuch für Industrie und Wissenschaft. Springer, Berlin 1984. p. 812.

[6.4] Jávor, A.: *Auswertungsmethode für die Bestimmung der integralen Nichtlinearität bei diskreten Meßwerten. Messen—Steuern—Regeln,* 12, 1967/7. pp. 266—267.

[6.5] *Mathematische Modellierung und Softwarenmässige Kompensation piezoresistiver Druckaufnehmer.* Technische Information Keller AG. für Druckmeßtechnik, Winterthur 1983.

CHAPTER 7

[7.1] Pattantyús, Á. G.: *Engineering Pocketbook.* (In Hungarian). Vol. I. Királyi Magyar Egyetemi Nyomda, Budapest 1937. p. 1567.

[7.2] Pattantyús, Á. G.: *A Handbook for Mechanical and Electric Engineers.* Vol. II. (In Hungarian). Practical Mathematics, Műszaki Könyvkiadó, Budapest 1961. pp. 455—562.

[7.3] Willers, Fr. A.: *Mathematische Instrumente.* Walter de Gruyter and Co., Berlin—Leipzig 1926. p. 144.

[7.4] Lochmann, G.: *Nomografie*. Fachbuchverlag, Leipzig 1979. p. 135.

[7.5] Székely, V. and Benkő, T.: *Characteristics, diagrams, nomograms*. (In Hungarian). Series Programming-electronics. Műszaki Könyvkiadó, Budapest 1975. p. 223.

[7.6] Levens, A. and Chalk, W.: *Graphics in Engineering Design*. (2nd Ed.) Wiley, New York 1980.

[7.7] Levens, A. S.: *Graphical Methods in Research*. Wiley, New York, 1965. p. 215.

[7.8] *Bildauswertsystem Videomat*. Druckschrift Nr. E 8606-T 6018-A 151-A-1, Siemens AG. Bereich Meß- und Prozeßtechnik, Abt. Bildtechnik, Karlsruhe

[7.9] Pühler, J.: *Störeinflüsse bei der Querkraftmessung am Balken*. Dissertation, Universität Karlsruhe 1987.

[7.10] Sorensen, H. W.: *Parameter Estimation. Principles and Problems*. Marcel Dekker Inc., New York and Basel 1980.

CHAPTER 8

[8.1] Bode, H. W.: *Network Analysis and Feedback Amplifier Design*. (In Hungarian). Princeton, Van Nostrand. 1945. Műszaki Könyvkiadó, Budapest 1961. p. 494.

[8.2] Wehrmann, W. *et al.*: *Real-Time Analysis*. Lexika, Grafenau 1979. p. 245.

[8.3] Géher, K.: *Linear Networks*. (In Hungarian). (2nd Ed.) Műszaki Könyvkiadó, Budapest 1972. p. 527.

[8.4] Géher, K.: *Theory of Network Tolerances*. Akadémiai Kiadó, Budapest 1971. p. 184.

CHAPTER 9

[9.1] August Fischer K. G. Göttingen edition: *Elektronisches Auswerten in der Meßtechnik*. p. 15. (1969)

[9.2] Lange, F. H.: *Korrelationstechnik*. (2nd Ed.) Verlag Technik, Berlin 1962.

[9.3] Wehrmann, W. *et al.*: *Korrelationstechnik*. (2nd Ed.) Expert, Grafenau 1980.

[9.4] DISA (works news): *Ein neuer Analog-Korrelator*. DISA Information. Herlev, No. 7. Jan. 1969. pp. 36–38.

[9.5] DISA (works news): *Amplitude Probability Analyser*. Herlev, DISA Information, No. 9. Feb. 1970. pp. 41–44.

[9.6] Wad, G.: *Amplitude Probability Analysis of Non-stationary Signals*. DISA Information, Herlev, No. 10, Oct. 1970. pp. 7–9.

[9.7] Jensen, J. O.: *A Survey of Methods for Measuring Correlation Functions*. Herlev, DISA Information No. 10, Oct. 1970. pp. 10–14.

[9.8] DISA (works news): *New Time Delay Unit Completes DISA Time Correlation Analyser* (TCA). DISA Information, Herlev, No. 9. Feb. 1970. pp. 37–40.

[9.9] Bendat, I. S. and Piersol, A. G.: *Measurement and Analysis of Random Data*. Wiley, New York 1966.

[9.10] Shu, W.: *Laufzeitkorrelation zur Geschwindigkeitsmessung mit Hilfe der diskreten Fouriertransformation*. Dissertation, Universität Karlsruhe 1987.

SUBJECT INDEX